Efficient Enforcement of Truth-Telling in the Grandfathering Process of an Emissions Trading Scheme

Kollektive Entscheidungen, Wirtschaftspolitik und öffentliche Finanzen

Herausgegeben von Cay Folkers

Band 15

PETER LANG
Frankfurt am Main · Berlin · Bern · Bruxelles · New York · Oxford · Wien

Frauke Eckermann

Efficient Enforcement of Truth-Telling in the Grandfathering Process of an Emissions Trading Scheme

PETER LANG
Internationaler Verlag der Wissenschaften

Bibliographic Information published by the Deutsche Nationalbibliothek
The Deutsche Nationalbibliothek lists this publication in the Deutsche Nationalbibliografie; detailed bibliographic data is available in the internet at <http://www.d-nb.de>.

Zugl.: Dortmund, Techn. Univ., Diss., 2008

Cover illustration:
Mehrgipfelige Präferenzen
und zyklische Mehrheiten.

Printed with kindly permission of
Prof. Dr. C. Folkers.

D 290
ISSN 1430-5852
ISBN 978-3-631-58402-6

www.peterlang.de

Vorwort des Herausgebers

Wirtschafts- und finanzpolitisches Handeln bedarf theoretischer und empirischer Fundierung. Es markiert die Schnittstelle zwischen dem ökonomischen und politischen Bereich und bedingt damit in zentraler Weise explizite Analysen kollektiver Entscheidungen. Ökonomische Analysen wirtschafts- und finanzpolitischer Fragen in ihrer Interdependenz mit politisch-ökonomischen Zusammenhängen kollektiver Entscheidungen sind die gemeinsame Klammer der in dieser Schriftenreihe veröffentlichten Untersuchungen. Der Herausgeber verfolgt das Anliegen, Beiträge zur anwendungsbezogenen Lösung theoretischer Grundfragen und zur theoriebasierten und empirisch gestützten Lösung praktischer Gestaltungsprobleme für wirtschafts- und finanzpolitische Fragen zur Diskussion zu stellen. Dabei sind Funktionsweise und Gestaltung politischer Institutionen ebenso als Untersuchungsgegenstände des Ökonomen anzusehen wie theoretische Fragen wirtschaftlicher Prozesse und öffentlicher Finanzen. Erst die gleichgewichtige und integrative Analyse ökonomischer und politischer Tauschprozesse führt zu kohärenten Beurteilungen und zu praktisch relevanten Empfehlungen für wirtschafts- und finanzpolitische Konzepte.

Die Reihe ist theorieorientiert unter besonderer Betonung der Analyse kollektiver Entscheidungen und anwendungsorientiert im Hinblick auf spezielle wirtschafts- und finanzpolitische Fragen im Zusammenhang mit den zugrundeliegenden politischen Verfahren und ihrer Reform. Damit ist ein Forschungsspektrum umschrieben, das ebenso allokationstheoretische, verteilungstheoretische und gesamtwirtschaftliche Problemstellungen in ihrem politischen Bezug wie Analysen von Institutionen und Entscheidungsprozessen in ökonomischer Perspektive umfaßt und die für Fragen der politischen Realität entscheidende Brücke zwischen ökonomischen und politischen Entscheidungen zu schlagen sucht. Der Fragenkreis schließt theoretische Analysen wirtschafts- und finanzpolitisch relevanter Problemstellungen, Untersuchungen zur Public Choice Theorie, zur Institutionenökonomik, zur ökonomischen

Analyse des Rechts und zur konstitutionellen Wirtschaftstheorie sowie verwandte Fragestellungen ein. Die Reihe ist für einschlägige Arbeiten offen und bemüht sich, die Vielfalt der theoretischen Ansätze und der praktischen Fragen in innovativen Untersuchungen zur Diskussion zu stellen. Sie will damit einen Beitrag zur Aufdeckung der Ursachen und zur Begründung von Lösungskonzepten für gesellschaftlich drängende Gegenwartsprobleme leisten.

Bochum, im Januar 1996

Cay Folkers

Preface

This dissertation draws on research I undertook during the time in which I held a scholarship of the Deutsche Forschungsgemeinschaft (DFG) at the Graduiertenkolleg "Allokationstheorie, Wirtschaftspolitik und kollektive Entscheidungen", and later while I was a teaching and research assistant at the chair of public finance at the Technische Universität Dortmund. The present thesis has strongly been influenced by and profited from discussions with professors and fellow students of the Graduiertenkolleg, as well as presentations at various national and international conferences. I am very grateful to all who supported my work in that way. Financial support by the Deutsche Forschungsgemeinschaft is gratefully acknowledged.

In particular, I would like to thank Wolfram Richter, who supervised my dissertation. Moreover I would like to thank Cay Folkers, Burkhard Hehenkamp, Wolfgang Leininger, Julia Angerhausen, Christoph Braun, Tobias Guse, and Christiane Schuppert, who all helped me to improve this thesis at the various stages of its development. Above all, I would like to thank my family for their strong support!

Dortmund, June 2008

Frauke Eckermann

Contents

List of Figures

List of Tables

Chapter 1

Introduction

Emissions trading as a policy instrument to achieve a predetermined environmental standard became famous in Europe with the start of the EU Emissions Trading Scheme in January 2005. In fact, this was the first time the instrument was applied in an international context, whereas the first national trading schemes were introduced in the United States in the 1970s. On a theoretical level, after COASE (1960) had cleared the way for this direction of thought, CROCKER (1966) and DALES (1968) were the first to propose a system of transferable property rights as a practical program for achieving given targets of pollution quantities of air and water respectively. More common to the economic profession are the articles by BAUMOL AND OATES (1971) and MONTGOMERY (1972), who established the formal theory and proof of cost-efficiency. Many challenges arise by putting such an instrument into practice. I analyze the problem of the necessary but costly high verification standard of the self-reported data and present cost-efficient ways to tackle it.

Emissions Trading

The past decades have not only seen an increasing awareness of a necessity for environmental protection and demand for political action, but also an increasing use of market based instruments for managing the environment. The consciousness of the problem of climate change was not least raised by the enhanced scientific evidence of the influence of emissions from economic activity on changes of the climate. A major channel for the publication of the present state of knowledge on climate change are the Assessment Reports of the Intergovernmental Panel on Climate Change (IPCC), which are based on peer reviewed and published scientific literature and serve as basis for political action[1]. The latest report, which is the fourth since 1990, came out in 2007 (IPCC, 2007), close in time to the publication of the Stern-Review (STERN, 2006), an analysis of the economic aspects of climate change on behalf of the British government. Both reports provide strong evidence of human-induced climate change and its potentially serious impacts and can be regarded as turning point in making the general public aware of the problem and its magnitude. Although it can not be stated precisely to how much global warming any particular increase in greenhouse gas concentration will lead, and how exactly that will affect the ecosystem in different regions of the world, scientists are much clearer about the potential impacts now than they have been years ago. Concerning the reality of anthropogenic climate change, there has long been agreement in the scientific community. A meta-analysis by ORESKES (2004) confirms this consensus: among the 928 analyzed papers, which were published in refereed scientific journals between 1993 and 2003, and listed in the ISI database with the keywords "global climate change", none disagreed with the hypothesis that the Earth's climate is affected by human activities. More uncertainties seem to have prevailed in the magnitude of the effects (MCKIBBIN AND WILCOXEN, 2002), which are important for the decision of costly emissions reduction today versus adaption to climate change in the future. Past Studies argued that among the two unknown cost components reduction and adaption, future costs are overestimated, while today's abatement costs are underestimated, and consequently gave the recommendation not to engage in too much emissions reduction today (NORDHAUS, 1993). The new reports, however, present different evidence on the magnitude of effects of climate change on ecosystems, as well as economic activity. Examining the economic impacts of climate change, STERN (2006) comes to the conclusion that the benefits of reducing emissions now will considerably outweigh the costs. One reason for the different results is the way uncertainties are treated. While earlier studies tended to exclude highly uncertain impacts of climate change, STERN (2006) explicitly includes these, arguing that recent scientific results have shown that the risk

[1] The IPCC was established in 1988 by the World Meteorological Organization and the United Nations Environment Programme, with the stated aim to evaluate the risk of climate change brought on by humans, see http://www.ipcc.ch/. It was awarded the Nobel Prize for Peace in 2007.

of outcomes much worse than expected is very real. Another reason for the different policy recommendations are different estimates for expected costs of emission reductions, which are lower in STERN (2006), as he puts more emphasize on improved and improving energy efficiency. The point that led to the largest discussions among the economic profession was the rate of discounting the future benefits and costs, that Stern used, which was criticized by many to be too low. STERNER AND PERSSON (2007), however, show, that results similar to those in the Stern Review can be obtained even without making the criticized assumption concerning the discount rate, when instead relative price changes are taken into account.

Having accepted the need for climate change policies, the question remains how to design them. To economists it is clear that the problem results from a market failure due to the presence of externalities, in this case the discharge of emissions without consideration of the global impacts. STERN (2006) for example calls climate change "the greatest and widest-ranging market failure ever seen." Among the many definitions of externalities, the one by ARROW (1969) (in HANLEY ET AL. (2007), p.49) seems to call for a solution of the problem via the creation of a market for emission rights: "a situation in which a private economy lacks sufficient incentives to create a potential market in some good, and the nonexistence of this market results in the loss of efficiency." Regarding the political action to be undertaken, STERN (2006) identifies emissions trading as one of the key elements of a necessary future international framework, and the National Center for Environmental Economics finds marketable permit systems to have "gained greater acceptance worldwide, particularly for the control of greenhouse gas emissions" (NCEE, 2004). This indicates that emissions trading seems to be on the rise in climate policies. As a reason why this happened not sooner, CUMBERLAND (1990), HANLEY ET AL. (1990) and STAVINS (1998) argue that economists might have put too much stress on efficiency in designing or evaluating an environmental policy, and neglected aspects such as distribution, interest group positions or scientific validity. In addition to these points, HAHN (1989) explicitly refers to monitoring and enforcement capacities as an important factor for the choice of instruments.

Since climate change is a global problem, it obviously calls for a global solution, which was answered with the Kyoto Protocol (UNFCCC, 1997).

The achievement of this treaty is the establishment of an international legal framework with binding emission targets. The environmental effectiveness, however, has to be questioned (McKibbin and Wilcoxen, 2002). The rules of the Kyoto Protocol provide three so-called *flexible instruments*, that aim to enable the reduction of emissions where it is least costly: Joint Implementation (JI), the Clean Development Mechanism (CDM) and International Emissions Trading (IET). JI and CDM are project based mechanisms, that allow a country to fulfill a part of its emission reduction obligations by financing specific projects which reduce emissions in another country. Under JI both participating countries are so-called Annex B countries, i.e. countries that are listed in Annex B to the Kyoto Protocol. These are mainly industrialized nations and economies in transition[2]. Under the CDM, Annex B countries can finance projects in non-Annex B countries, which are mainly developing countries. A more detailed description of these two instruments can be found in Michaelowa (2004). The foreseen Emissions Trading Scheme is regulated by Article 17 of the Kyoto Protocol (UNFCCC, 1997). It allows Annex B countries the fulfilment of their national quotas via trading of emission rights with other Annex B countries. The trading scheme under the Kyoto Protocol differs from other trading schemes insofar as it foresees trading among countries and not companies, and as the initial allocation of allowances for each country results from political negotiations. Still, it was a major step towards the application of this instrument on an international level.

Existing Trading Schemes

In the meantime there is a multitude of experiences with this instrument, most of it in the United States, showing constant learning from past programs and the achievement of policy goals at significantly lower costs than under command-and-control alternatives in some cases (Stavins (2001), Hahn and Hester (1989b)). An overview of the evolution of emissions trading schemes is provided in Tietenberg (2006). A study by the United Nations (UN, 2001) discusses national trading programs in Europe, that emerged between the ratification of the Kyoto Protocol and the start of the EU Emissions Trading Scheme.

[2] The Annex B to the Kyoto Protocol is imprinted in Appendix A of this analysis.

One of the most successful trading schemes is the *Sulphur Trading Program*, which was established in the United States by the Environmental Protection Agency (EPA) in 1990 and aimed to cut emissions of sulphur dioxide (SO_2). It was successful with respect to the achievement of targeted emissions reductions (which it exceeded), low abatement costs and traded volume (STAVINS (1998), BURTRAW AND PALMER (2004) and others). The success of this program lead to an increased acceptance of the instrument. It was to a great extend due to the fact that the EPA organized a trading platform, the so-called auction market, by which transaction costs could be reduced to a large extend (SCHMALENSEE ET AL., 1998). Another important factor were the strict penalties for noncompliance (STAVINS, 1998) that lead to a very high degree of compliance and market confidence.

An example where the concept of tradeable property rights has been applied successfully to an input to emissions, not emissions themselves, was the *Lead Phase-out Program*, designed for the establishment of the production of gasoline without lead. Another difference of this program was that its aim was not only to meet a certain cap, but to eliminate the pollutant completely. The drivers for the success were that (a) refineries were the actors, not car-holders, which obviously reduced monitoring and enforcement efforts, and (b) the fact that refineries could receive and trade rights by early elimination of lead. The consequence was an earlier phase-out than under direct regulations, since these would have lacked the incentive for the refineries to act early (TIETENBERG, 2006).

The *EU Emissions Trading Scheme* then had the "second mover advantage" in being able to include lessons learned from other programs into its design. While STAVINS (1998) coined the expression "Grand Policy Experiment" for the Sulphur Trading Program, KRUGER AND PIZER (2004) judge the EU Emissions Trading Scheme to be the "New Grand Policy Experiment", due to its size, scope and multi-jurisdictional political structure. The EU Emissions Trading Scheme is the core instrument for Kyoto compliance in the EU. Its aim is the reduction of greenhouse gas emissions in order to meet the Kyoto targets, which constitute a reduction of 8% of 1990 emissions over the period from 2008 to 2012 for the European Union (cf. UNFCCC (1997), Figure A-1 in Appendix A of this analysis). Emissions trading as an instrument was chosen, since several studies have shown great cost advantages compared to scenarios with no trading, see EC (2000) and the literature therein. The EU Emissions Trading Scheme (EC, 2003) started in January 2005 with a first trading phase from 2005 to the end of 2007, that was often referred to as test phase. The second trading phase of the EU Emissions Trading Scheme will coincide with the commitment period under the Kyoto Protocol (from 2008 to 2012) and will be followed by further trading phases. The launch of the trading scheme before the legally binding phase under the Kyoto Protocol enables European companies (and countries) to get acquainted with the instrument and, if necessary, allows the European Commission to make adjustments in time. Traded units are the so called greenhouse gas allowances, which are measured in tons of CO_2 equivalents[3]. Participation in the trading program is mandatory for large installations of energy-intensive industries of all 25 EU member states and trading takes place among firms of the same as well as different branches and countries. After initial resistance from some industries, the scheme has now found general acceptance. Evaluations of the first trading phase by STERN (2006) and MCKINSEY AND ECOFYS (2005) highlight amongst others the importance of long term confidence in the future of the scheme and the necessity to reconsider the costs that arise in particular for small firms from the compliance standards. As the major part of participating firms is relatively small (50% of the installations received less than 2% of emissions allowances in the first trading period, cf. Figures A-2 to A-4 in Appendix A), this problem is not negligible. Among the German firms that take part in the EU Emissions Trading Scheme, for example, the 30 biggest firms account for nearly 50% of emissions, while the 1000 smallest firms only contribute approximately 2% of emissions, see Figures A-5 and A-6 in Appendix A.

Implementation: Crucial Factors

Emissions trading is a cap and trade instrument: a given amount of property rights is distributed to firms, which can then trade these rights. Thus, firms

[3] There are six greenhouse gases which are addressed in the Kyoto Protocol: carbon dioxide (CO_2), methane (CH_4), nitrous oxide (N_2O), hydrofluorocarbons (HFCs), perfluorocarbons (PFCs) and sulphur hexafluoride (SF_6). The gases are compared by transforming them into CO_2 equivalents. REILLY ET AL. (2000) have shown that the inclusion of all greenhouse gases leads to cost savings compared to the case where the Kyoto targets are reached by reducing CO_2 alone.

are free to decide how many emissions to abate and how many emission rights to buy or sell. Two challenges related to the implementation of this instrument can directly be seen: the initial allocation of allowances and the establishment of the respective "financial" market. A third element, however, is just as important: monitoring and verification of all firm data in order to establish confidence in the system.

In the EU Emissions Trading Scheme the allocation of allowances is carried out on a national level. Each country holds a budget of emission allowances, which is determined by its commitments under the Kyoto Protocol. For each trading period, countries set up their National Allocation Plans (NAPs), in which they lay down how their allowances are distributed among the installations in their country. These plans have to be approved by the European Commission. They have to fulfill the requirement that the allocation of allowances to the participating firms must still ensure the compliance of the Kyoto targets. In particular it has to be considered that there will also be emissions from other sectors and companies that do not take part in the trading scheme (EC (2003), Annex III). The regulations of the NAPs will also determine the scarcity in the market. Experience from the first trading period shows that allocations have not been strict enough, and consequentially the price for allowances has dropped heavily after actual emissions of the participating installations were released (STERN (2006) and others).

With respect to the actual allocation, three methods are under discussion: grandfathering, benchmarking and auctioning. Under grandfathering and benchmarking, firms receive the allowances free of charge, based on historic emissions in case of grandfathering and on technical performance in case of benchmarking. Under auctioning, firms can obtain allowances in an auction, where the type of the auction is not prescribed in the guidelines. A vast deal of literature has evolved on the question which allocation method should be preferred. There is a general consensus in the economic literature that auctioning should be preferred to grandfathering or benchmarking (GRAICHEN AND REQUATE (2003), CRAMTON AND KERR (2002), JENSEN AND RASMUSSEN (2000) and others), as the latter may lead to strategic behavior of and "windfall profits" for firms, and require a higher administrative effort. Auctions, on the other hand, raise revenues and provide an equal opportunity for new entrants. MONTERO (2008) has shown that an auction with paybacks can even be used to establish a first-best solution. KRUGER AND PIZER (2004) argue, that the reason why, despite of the theoretic evidence, so few auctions can be seen in existing trading programs, is a political one. For the US Sulphur Trading Program, for example, this is confirmed by JOSKOW AND SCHMALENSEE (1998). In the EU Emissions Trading Scheme nearly all allowances are grandfathered free of charge. For the first trading period the EU Directive allows a maximum of 5% of allowances to be auctioned, and for the second trading period a maximum of 10% (EC, 2003), but this option was only used by few countries.

Once allowances are allocated and stock exchanges are established, two aspects are, from the economic literature, known to be crucial for the functioning of the system: market power (HAHN (1984), LISKI AND MONTERO (2005)) and transaction costs (STAVINS (1995), MONTERO (1997)). Concerning transaction costs, KERR AND MARÉ (1998) support theoretical results with an empirical study of the US Lead Phase-Out Program, finding that, although in general this market worked very well, agents facing high transaction costs suffered significant efficiency losses. GANGADHARAN (2000) analyzes another permit market, launched in Los Angeles in 1994: the Regional Clean Air Incentives Market (RECLAIM) and finds that transaction costs were substantial in the initial years of the program. Regarding market power, however, there does not seem to be strong empirical evidence (HANLEY ET AL., 2007). This experience seems to be reflected also in the EU Emissions Trading Scheme: while there is to date no evidence that market power seems to have been a serious problem, transaction costs are large for small firms (EGENHOFER AND FUJIWARA, 2005). Other important issues in this scheme turned out to be the rules for linking the scheme with other trading schemes and for the inclusion of allowances that were obtained through the use of the Clean Development Mechanism (STERN, 2006).

Initial allowances are allocated based on firms' reports of their historical emissions, additional allowances can be obtained via projects in another country - fundamentals of the scheme that rely on self-reported data. Needless to say that strict verification is obligatory. The guidelines concerning "monitoring, reporting, verification and compliance" are laid out in an extra document (EC, 2004), stating the importance of a high level of compliance in every country for establishing international confidence in the scheme and meeting the environmental target. In particular, the EU Directive (EC, 2003) requires an operator of an installation to show, that the reported data is free of inconsistencies, that the collection of the data has been carried out in accordance with the applicable scientific standards, and that the relevant records of the installation are complete and consistent.

In his evaluation of the US Sulphur Trading Program, STAVINS (1998) points out that the strict penalties that were applied were a main reason for the high degree of compliance, but also that the costs combined with the monitoring that was necessary for the enforcement were significant. In the EU Emissions Trading Scheme, penalties are as well used for non-compliance. KRUGER AND EGENHOFER (2006), however, advise to introduce penalties also for firms that falsify information. The EU-Directive leaves the decision on penalties for fraudulent reporting to member states (EC, 2003). KRUGER AND PIZER (2004) argue that this may lead to non-uniform penalties, since they will depend on national laws. So far, however, countries make no use of the option to punish false reports. The German guidelines, for example, foresee only an upgrading of incorrect data (UMWELTBUNDESAMT, 2004).

Information Revelation

This study analyzes how, by the introduction of penalties for incorrect reports, audit costs may be reduced in the process of the initial allocation of emission allowances. Other fields of literature where the verification of private data plays a crucial role are the fields of compliance and enforcement, crime and punishment, and tax evasion. At first sight, the verification of firm data in an emissions trading scheme seems to belong to the field of *compliance and enforcement* problems[4], and in particular be related to models with self-reporting of the compliance status, where the inspection depends on these reports (as for example in MALIK (1993) and HARFORD (1987). There is, however, an important difference to the verification problem which is analyzed in this study. I do not consider compliance problems, but problems related to the reporting of the data. While compliance imposes costs on firms, they incur no costs for reporting false data unless they are audited and penalized. In addition, the compliance literature very often focusses on

[4] An overview over the extensive literature on this is given by HEYES (2000).

the maximization of social welfare, not the explicit establishment of high verification standards.

COHEN (2000) points out the strong relation between environmental enforcement and the *crime and punishment* literature. The model of KAPLOW AND SHAVELL (1994), for example, is close to the ones mentioned above. They analyze the effects that self-reporting in legal enforcement has on social welfare. In a model where individuals may or may not commit a harmful act they find that schemes with self-reporting are advantageous, since investigation costs can be saved when individuals report their harmful behavior. Naturally, the sanction for the reported harmful act is equal to or smaller than the expected fine when the act is not reported. The literature of the economics of crime started with the seminal article by BECKER (1968). A comprehensive summary of this theory is provided by POLINSKY AND SHAVELL (2000), in which environmental enforcement is listed as an example for the application of public enforcement of law. The verification of firm data in the grandfathering process of an emissions trading scheme fits into this strand of literature insofar as a false report can be regarded as an offence and audit costs can be reduced by increasing the penalty - or in the context of this analysis, by introducing a penalty. The difference is the objective of the regulator, which is in the crime literature, as in the compliance literature, the maximization of social welfare, i.e. the determination of an "optimal" level of enforcement, and not the implementation of high verification standards, i.e. the reduction of offences to zero.

Thus, the objective of this analysis is more in line with that of the *tax evasion* literature, where many studies make use of the revelation principle and, among the class of incentive compatible income tax and enforcement schemes, characterize the optimal ones. While the first study on the problem of tax evasion focussed on the taxpayer's behavior (ALLINGHAM AND SANDMO, 1972), the survey by SANDMO (2005) shows that in the years since, questions have been expanded to topics such as evasion and avoidance, the connection between tax evasion and labor supply, evasion by firms, and, not least, the impacts on optimal taxation. Just as in the MIRRLEES (1971) study, the consideration of tax evasion leads to an equity-efficiency (here "audit-cost-efficiency") tradeoff. It arises, since in order to induce truthful reporting the optimal tax schedule must be regressive, see for example CHANDER AND

WILDE (1998) or MARHUENDA AND ORTUNO-ORTÍN (1997). This result can also be found in a setting where a distinction is made between the tax authorities (the government) and the tax enforcement agency, as in CREMER ET AL. (1990). Another common result is the necessity to reward honest reporting in order to induce truth-telling cost-efficiently (BORDER AND SOBEL (1987), MOOKHERJEE AND PNG (1989) and others). BOADWAY AND SATO (2000) confirm this result in a context with unintentional compliance errors by taxpayers and administrative errors by tax auditors. In addition they find that taxpayers report truthfully, so that the only "tax evaders" are innocent ones. Thus, the government's objective to redistribute is limited.

The problem of the verification of private data in the grandfathering context seems to be very close to that of tax evasion. There are, however, substantial differences which will be highlighted below. The most apparent is the fact that in the tax evasion context firms (or individuals) have an incentive to *understate* their private income, while in the grandfathering process of an emissions trading scheme firms have an incentive to *overstate* their emissions.

The literature on the verification of self-reported firm data in emissions trading schemes to date mainly concentrates on technical or legal issues, not on economic efficiency.

This Analysis

The grandfathering process of an emissions trading scheme provides an example of informational asymmetries with incentives for the firms to overstate the private data. In the EU Emissions Trading Scheme, for example, grandfathering was the allocation method for the first phase and will also be the allocation method for the second phase of the trading scheme, that is starting in 2008. Allocated allowances will then be based on an average of the emissions of the years 2000 to 2005. In addition, countries like Australia and an initiative of several states in the USA[5] are planning to introduce a national trading scheme and link it to the EU Emissions Trading Scheme. Strict verification of the self-reported data is essential but costly, which raises the question of cost efficiency, that is discussed in this analysis. The motivation for the topic can be summarized by the statement by KRUGER AND EGENHOFER (2006), page 9: "Consistent and fraud-proof monitoring and

[5] Regional Greenhouse Gas Initiative (RGGI) www.rggi.org

verification procedures therefore have moved to the centre of emissions trading schemes. While this debate initially focussed on accuracy and credibility, with progressive implementation the spotlight turns to costs, notably how to reduce them."

The problem of the verification of self-reported emissions data for the initial allocation of emission allowances resembles the one of tax evasion, where firms have to pay a tax based on their (voluntarily) reported income. Following the approach that Chander and Wilde (1998) have used for the characterization of optimal income tax and enforcement schemes, I set up a principal-agent model with costly auditing, exogenously constrained penalties, and risk neutral firms and government. Verification schemes are introduced which consist of three elements: Firstly, emission allowances are allocated on the basis of the reports of the firms via an allocation function. Secondly, audit probabilities are determined for each firm, based as well on the reported data. Thirdly, penalties are imposed for overstatements. The model is static and it is assumed that there is no commitment problem. The politically desired high verification standard is given by incentive compatibility in the model.

Only few requirements are necessary for the elements of the verification schemes, most of them arising from the political context. Two, however, have to be highlighted, since they, while being reasonable and straightforward, make additional conditions necessary for the use of the revelation principle: firstly the assumption that only firms with a positive amount of emissions will receive any allowances and secondly the assumption that there is no reward for underreporting or truth-telling. Together with the fact that in this model firms have an incentive to overstate their emissions, while for tax evasion there is an incentive to understate the income, these two assumptions form major differences to the tax evasion literature. Another aspect that arises in particular for this problem is the exogenously given overall amount of emission allowances that may be allocated, due to which the problem may seem to be a rent-seeking problem. It is shown, however, that firms do not take their rivals' actions into consideration, and that therefore the principal-agent framework is appropriate.

In Chapter 2 of this analysis the model is presented, as well as a general characterization of efficient verification schemes, where an efficient scheme has the property, that, for a given penalty structure, incentive compatibility can not be achieved with less auditing. I derive and discuss the interplay between the allocation function, audit probability and penalty structure in efficient verification schemes. As expected, the "Becker solution" holds in the general model-setup and audit costs can be reduced by increasing the penalty. Too high penalties are, however, likely to be politically not feasible, and therefore the efficiency condition is further specified for alternative penalty, as well as audit structures.

The analysis in chapter 3 points out how the efficiency results, and with this the policy recommendations, depend on the different penalty structures, and in how far the request of incentive compatibility is restrictive. While the performed analysis does not allow, and not intend, a direct ranking of the different penalty structures, one that combines the gain from overstating with an additional penalty part seems to be most advantageous. If the intention had been a direct ranking of penalty structures, it would surely have been a ranking with respect to their potential to reduce audit costs, since the overall aim is the reduction of these costs. It is immediately seen, however, that then an infinitely high penalty should be picked, which, as noted above, is not feasible. This explains the focus of the analysis on cost-efficiency for a *given* penalty structure. Even a definite audit of every report must be called cost-efficient under a respective penalty, although it certainly is not the least costly approach. While the efficiency results depend strongly on the underlying penalty structure, they do not depend on the distribution of emissions across firms. In chapter 3 it is also set out how the differences to the setup of the related field of tax evasion find their counterpart in the differences in the results.

In chapter 4 an audit structure is analyzed that provides a straightforward method of complying with a given audit budget: all firms that report emission values above a certain cutoff-level are audited, while the others are not. Starting from a scheme with a definite audit for every report, a reduction of audit costs via such a deterministic audit structure seems to be less complex than a reduction of costs via the random audit method analyzed in chapter 3, which requires an accurate consideration of the interrelation of audit probabilities, allocation and penalty structures. It is shown, however, that the price for this "simplification" is a redistribution of emission allowances from firms with an emissions value below the cutoff-level to those with emissions above the cutoff-level. A comparison of total audit costs shows, that they are not per se smaller for any of the audit schemes. They may be higher for random or deterministic auditing, depending on the allocation function (which determines the audit probability for efficient random audits), the distribution of emission values across firms, and the cutoff-level.

Conclusions are provided in chapter 5 with a focus on policy recommendations. Major proofs are relegated to the appendices.

Chapter 2

Analytical Framework

Since the initial allocation of greenhouse gas allowances to firms via grandfathering is primarily based on the firms' own emissions reports, it is commonly agreed that a thorough verification of these reports is necessary. The question is just how thorough this verification should be, whether it should be combined with penalties, and what the structure of such verification schemes should be. In this chapter I set up the principal-agent model and present the general results on which the following efficiency analysis is based.

2.1 Verification Mechanisms

In the model the true emission value $\theta \in [0, \theta_u]$ of a firm is given exogenously and considered to be private information of the firm. Emission allowances are then grandfathered to the firm, solely based on its own report of its emissions. The model sets in at the point where the firm reports an emission value $\hat{\theta} \in [0, \theta_u]$ to the government. Since the firm may have incentives to report a value $\hat{\theta} \neq \theta$, the government may prefer to perform costly audits and to impose penalties for misreporting. In this principal-agent framework it is assumed that the principal and agent are risk-neutral and that there is no commitment problem.

The government introduces a *verification scheme*[1] $(e(\cdot), a(\cdot), f(\cdot))$, which consists of the *allocation function*, which depends on the firm's report $\hat{\theta}$ and

[1] Technically correct the verification scheme is a mechanism that is given by $\left(\hat{\Theta}, (e(\cdot), a(\cdot), f(\cdot))\right)$, where $\hat{\Theta}$ is the message space (reported emission types) of the firms and $(e(\cdot), a(\cdot), f(\cdot))$ the outcome function. However, as the outcome function determines the mechanism, the introduced notation is used.

determines the amount of *emission allowances* the firm shall receive,

$$e : [0, \theta_u] \rightarrow \mathbb{R}_+, \tag{2.1}$$

the *audit probability*, which also depends on the firm's report $\hat{\theta}$,

$$a : [0, \theta_u] \rightarrow [0, 1], \tag{2.2}$$

and the *penalty function* or *fine*, which depends on the firm's report $\hat{\theta}$ as well as the true emission value θ,

$$f : [0, \theta_u] \times [0, \theta_u] \rightarrow \mathbb{R}_+. \tag{2.3}$$

Initial allocations of emission allowances, as well as fines, are non-negative. All functions are assumed to be continuous, but differentiability is not required. The continuity-assumption will be relaxed later on in some parts of the analysis. Thus, a firm which reports $\hat{\theta}$ receives an initial amount of emission allowances $e(\hat{\theta})$. In addition, it will be audited with probability[2] $a(\hat{\theta})$ and will then have to pay the fine $f(\hat{\theta}, \theta)$. It is assumed, that the true parameter is discovered, if a firm is audited. In addition to the assumption that it is technically possible to observe the true emission value, this includes the assumption that bribing does not take place.

It is assumed that the determination of the audit probability as well as the payment of the audit costs are responsibilities of the government. The placing of the burden of the actual audit costs onto the side of the government does not conflict with the perception that audit costs should be paid by firms. In such a case the underlying assumption of lump sum payments from firms to the government holds. This concept is not new to emissions trading. Under the EU emissions trading scheme, for example, it is being used by the German Emissions Trading Authority (DEHSt) to cover costs for administrative responsibilities.

In this model the penalty also consists of emission allowances and represents the amount of allowances that is transferred from the firm to the government. *Expected allowances* for the firm are then given by initial allocations minus expected penalties:

$$\pi(\hat{\theta}, \theta) = e(\hat{\theta}) - a(\hat{\theta}) f(\hat{\theta}, \theta). \tag{2.4}$$

[2] As the firm will only be audited with a certain probability, one might see the firm's problem as a decision in a lottery between a risky and a risk-free payoff.

Since costs are lump sum and since allowances are grandfathered free of charge, it is the objective of the firm to choose $\hat{\theta}$ such that expected allowances are maximized.

The objective of the government is to construct the verification mechanism in an optimal way, where optimality is characterized by incentive compatibility and efficiency, and abstracts from redistributive considerations. The single components of the mechanism are required to fulfill certain social norms, which will lead to a set of feasible verification schemes which is displayed below.

The choice of the penalty function plays a crucial role in the analysis. To start off, I consider the penalty function to be a choice variable of the planner, subject only to slight restrictions resulting from the discussions in the political context. It should provide neither a reward nor a fine for understating or truth-telling, i.e. $f(\hat{\theta}, \theta) = 0$ should hold for $\hat{\theta} \leq \theta$. This implies that honest reports are not treated differently depending on whether they are audited or not. In addition, the penalty should "fit the crime", i.e. the penalty function should be bounded, and it should be nondecreasing in the degree of misreporting.

Truthful reporting will always lead to nonnegative expected allowances, however, (2.4) may become negative if $\hat{\theta} > \theta$ and the allocation function is small, or the expected penalty high enough. Since allocations and penalties are measured in allowances and since firms cannot give more allowances than they receive, it is assumed that "negative allowances" can be transferred into monetary equivalents in order to assure that the penalty is credible.

The *feasibility conditions* are given by:

$$e(0) = 0, \tag{2.5}$$

$$e(\cdot) \not\equiv 0, \tag{2.6}$$

$$f(\hat{\theta}, \theta) = 0 \quad \forall \hat{\theta} \leq \theta, \tag{2.7}$$

$$f(\hat{\theta}, \theta) \leq B \text{ monotonically increasing in } \hat{\theta} \text{ and} \tag{2.8}$$

$$f(\hat{\theta}, \theta) \leq B \text{ monotonically decreasing in } \theta \quad \forall \hat{\theta} > \theta, \tag{2.9}$$

$$\text{with } B \in \mathbb{R}_+, \ B < \infty.$$

A firm which reports zero emissions will not receive any allowances, which is given by equation (2.5). Equation (2.6) simply technically rules out the possibility of not allocating any allowances. (2.7) and (2.8) follow from above.

In addition, two technical conditions are introduced, which are necessary to make use of the revelation principle. In the following section 2.2 I will elaborate on the desirability of the application of the revelation principle, and of the necessity of these conditions:

$$\pi(\hat{\theta}, 0) \leq 0 \qquad \forall\, \hat{\theta} \in [0, \theta_u], \tag{2.10}$$

$$f(x, z) \leq f(x, y) + f(y, z) \quad \forall\, x, y, z \in [0, \theta_u] \quad \text{with } x \geq y > z. \tag{2.11}$$

These restrictions might be considered truly restrictive. Since they stem from the necessity to induce incentive compatibility the question arises, how restrictive this requirement is. The question will be answered in several subsections in the analysis below. Inequality (2.10) ensures, that a firm with zero emissions has no incentive to overstate, since it does not expect to receive any allowances, i.e. incentive compatibility holds for firms of type zero. Together with $a(\cdot) \in [0, 1]$ this implies

$$e(\hat{\theta}) \leq a(\hat{\theta}) f(\hat{\theta}, 0) \leq f(\hat{\theta}, 0) \quad \forall\, \hat{\theta} \in [0, \theta_u]. \tag{2.12}$$

For penalty functions that can be considered as distance functions, i.e. that solely depend on the difference between the true and the reported parameter, $f(\hat{\theta}, \theta) = f(\hat{\theta} - \theta)$, condition (2.11) denotes subadditivity. Let $b := (x - y)$ and $c := (y - z)$, then (2.11) is equivalent to $f(b + c) \leq f(b) + f(c)$. Thus, the penalty function has the property that the fine for a single "large" deviation is small relative to the fine of other combinations of "smaller" deviations that add up to the large deviation. In other words: penalties can't be reduced by splitting overstatements into parts. Such a penalty might restrain firms from splitting their plants and making overstatements in the single reports[3]. This restriction on the set of feasible penalty functions is further discussed below, together with the considerations leading to the revelation principle equivalent.

The set F of *feasible verification mechanisms* is given by:

$$F = \{(e, a, f) : \quad (2.5) \text{ to } (2.11) \text{ hold}\}. \tag{2.13}$$

[3] The importance of the property of subadditivity is well known from the theory of natural monopoly, where a single firm can produce the desired output at lower cost than any combination of two or more firms, see SHARKEY (1982), or from insurance mathematics, where the principle of premium calculation fulfills subadditivity in order to ensure that insurants cannot reduce their premium by splitting their risk into partial risks, see VYLDER ET AL. (1984).

One may argue, that the problem of the allocation of the initial allowances to firms is not a principle-agent problem, since there are many firms and since firms know that there is an upper limit of allowances and thus act strategically and take their rivals' actions into consideration. Two explanations, however, justify the choice of the principle-agent framework. The intuitive explanation is that each firm receives an amount of greenhouse gas emission allowances that is small relative to the number of total allowances. The other is a technical explanation. Suppose firms considered that there is a cap on allowances and that the other firms' decisions have an influence on how much they get and vice versa. This would result in a rent-seeking contest where players compete for emission allowances. It can then be shown, however, that it is optimal for firms not to take their rivals' actions into consideration. I provide the corresponding rent-seeking model and show the result in chapter 2.6. The result is driven by the fact that the penalty function $f(\cdot)$ does not depend on the other firms' actions. This assumption can be fully justified by either political considerations or by comparisons with literature on tax evasion or crime and punishment, where usually such a penalty is applied.

In some parts of the analysis it will be necessary to consider the entire firms in the economy. The set-up for this is as follows: In the model there are n firms, indexed by i, which are distinguished only by their emission values. The emission values θ are distributed over the interval $[0, \theta_u]$ according to a common probability density function $g : [0, \theta_u] \rightarrow \mathbb{R}_+$. The government knows the probability distribution of the emission parameters, but not the realizations of specific firms. The actual costs of auditing a firm are assumed to be equal for all firms and can therefore be normalized to one.

There is a cap on overall allowances, which I call V. If the sum over all firms' initial allowances exceeds this bound, it is assumed that they can be adjusted proportionally via a compliance factor. Such a procedure was proposed in the context of the EU Emissions Trading Scheme by HARRISON AND RADOV (2002) and has actually been applied in Germany by the name of "2nd capping". Further specifications of the determination of the initial allowances and the compliance factor are discussed below.

2.2 Incentive Compatibility

All mechanisms in F are *direct revelation mechanisms*, since firms are asked to report their emissions parameter and therefore for each firm the message space coincides with the set of possible types $[0, \theta_u]$. A direct revelation mechanism is said to be *incentive compatible* if it is optimal for each firm to report its emission value truthfully[4]. In the context of this analysis firms will report truthfully if this maximizes their expected allowances i.e. if

$$\pi(\theta, \theta) \geq \pi(\hat{\theta}, \theta) \quad \forall \hat{\theta}, \theta. \tag{2.14}$$

The government performs costly audits in order to induce truth-telling by the firms. I make the assumption that it is considered politically desirable that *all* firms report their emissions truthfully. Economically, however, the restriction of the analysis to incentive compatible mechanisms can only be approved if it is assured that no, possibly optimal, outcome is lost. The tool that is usually applied to ensure this, is the *revelation principle.* It is applied to a wide range of problems and states that without loss of generality the government can restrict its attention to incentive compatible direct revelation mechanisms. Or in other words: the government can implement a policy that enforces truth-telling, without losing optimality. The expression *revelation principle* appeared first in MYERSON (1981), in the context of auction design problems. However, the result was already established earlier for the implementation of social choice rules by DASGUPTA ET AL. (1979), or in the context of Bayesian collective choice problems by MYERSON (1979), and first seminal works go back to GIBBARD (1973) and HURWICZ (1973). The revelation principle finds its way into the regulation literature in BARON AND MYERSON (1982) and others. HARRIS AND TOWNSEND (1981) state the result in their characterization of efficient allocation processes for environments with asymmetric information, providing a principal-agent environment as an example. An introduction into the use of the revelation principle in principal-agent models is provided by LAFFONT AND MARTIMORT (2002). GREEN AND LAFFONT (1986) point out that the revelation principle need not apply in a principal-agent model with adverse selection, if the set of reports available to the agent depends on his type. They provide an additional condition, under which an equivalence to the revelation principle holds.

[4] Definitions are given e.g. in MAS-COLELL ET AL. (1995) or KREPS (1990) .

In the context I consider, the standard arguments do not apply due to requirements that originate from political considerations. I explicitly rule out rewards for truthful reporting, whereas in the tax evasion literature several authors establish the result that the optimal solution requires that there should be rewards for honest reporting (see MOOKHERJEE AND PNG (1989), CREMER AND GAHVARI (1996), and BOADWAY AND SATO (2000)).

BOADWAY AND SATO (2000) also point out, that in the absence of rewards for truthful reporting there is no guarantee that the revelation principle will apply. In my model, feasibility conditions (2.5) and (2.7) fix the image of the allocation- and the penalty function respectively at the point zero and make additional conditions necessary. Without conditions (2.5) and (2.7) it is easily seen that the revelation principle holds: if the mechanism $(e(\cdot), a(\cdot), f(\cdot))$ results in an optimal report $\alpha(\theta)$ for an emission value θ, the mechanism given by $(e'(\hat{\theta}), a'(\hat{\theta}), f'(\hat{\theta}, \theta)) := (e(\alpha(\hat{\theta})), a(\alpha(\hat{\theta})), f(\alpha(\hat{\theta}), \theta))$ has an optimal report which is equal to the true value θ. The payoffs to both the government and the firm are unaffected by this change, since $a'(\theta) = a(\alpha(\theta))$ and $\pi'(\theta, \theta) = e'(\theta) - a'(\theta)f'(\theta, \theta) = e(\alpha(\theta)) - a(\alpha(\theta))f(\alpha(\theta), \theta) = \pi(\alpha(\theta), \theta)$. Condition (2.7) requires that $f(\hat{\theta}, \theta) = 0$ for all $\hat{\theta} \leq \theta$, for reasons explained above. However, under standard arguments this is violated by the new mechanism (e', a', f'), since (e, a, f) is not incentive compatible, i.e. $\alpha(\theta) \neq \theta$ for some θ, and therefore $f'(\theta, \theta) := f(\alpha(\theta), \theta) \neq 0$ cannot be ruled out. Similar arguments apply to feasibility condition (2.5): $e'(0) = e(\alpha(0))$ need not necessarily be zero. This shows that the mechanism $(e'(\hat{\theta}), a'(\hat{\theta}), f'(\hat{\theta}, \theta)) := (e(\alpha(\hat{\theta})), a(\alpha(\hat{\theta})), f(\alpha(\hat{\theta}), \theta))$ cannot be used to prove the revelation principle in this setting. A different mechanism is therefore put up in the proof of Lemma 1 in Appendix B. It will also come clear from this proof that (2.10) and (2.11) are sufficient to ensure that without loss of generality the analysis can be restricted to incentive compatible mechanisms in F.

Lemma 1 *Let $(e, a, f) \in F$. Then there exists an incentive compatible direct revelation mechanism $(e', a', f') \in F$ such that e' is nondecreasing and (e', a', f') replicates the equilibrium outcome arising from (e, a, f).*

The proof is given in Appendix B. Note that $e(\cdot)$ nondecreasing is not an assumption, but an implication of the model. Another implication of the setup is the already mentioned restriction for the penalty function. In addition, assigning no allowances to firms with no emissions doesn't come at no costs, since it requires that incentive compatibility is provided for these firms:

Remark 1 *1. If, in contrast to general results in for example the tax evasion literature, it is (politically) desired that truthful reports are* not *rewarded, it is necessary to put an additional restriction on the set of feasible penalty functions (condition (2.11)), in order to obtain incentive compatibility.*

2. Likewise, for the additional request $e(0) = 0$ it becomes necessary to demand in addition incentive compatibility for firms of type zero (condition (2.10)).

Thus, the government should choose the verification mechanisms such that not only the politically straightforward conditions (2.5) to (2.8) hold, but also the, at first sight less intuitive, conditions (2.10) and (2.11). This restriction excludes, however, some penalty functions that might seem desirable. I will come back to this point in chapter 2.5 on penalty functions.

It is assumed that firms report their emission parameter truthfully if reporting truthfully is optimal. Due to Lemma 1 the analysis can then be restricted to mechanisms $(e, a, f) \in F$ which satisfy:

1. $e(\cdot)$ is nondecreasing:

$$\hat{\theta}^i > \hat{\theta}^j \Rightarrow e(\hat{\theta}^i) \geq e(\hat{\theta}^j), \tag{2.15}$$

2. incentive compatibility holds:

$$e(\theta) \geq e(\hat{\theta}) - a(\hat{\theta}) f(\hat{\theta}, \theta) \quad \forall \hat{\theta}, \forall \theta. \tag{2.16}$$

Denote this subset of mechanisms with F_I:

$$F_I = \{(e, a, f) : (e, a, f) \in F \text{ and (2.15) and (2.16) hold}\}. \tag{2.17}$$

Conditions (2.15) and (2.16) imply that $a(\cdot) \not\equiv 0$, unless $e(\cdot) \equiv const.$ So, whenever the allocation of allowances is differentiated with respect to the emissions of the firms (which of course is the intention), some auditing is necessary to achieve incentive compatibility.

2.3 Efficiency

High verification standards cannot be obtained without costs. In this model the desired standard is represented by incentive compatibility and the costs that have to be paid for it are given by the audit costs. In addition to incentive compatibility it is the objective of the government to allocate the allowances efficiently. It will come clear in the following section, that the so called "Becker conundrum" applies when penalties are determined endogenously - it is optimal to set the fine at its maximal level and minimize the costly auditing. To prevent this, since it is hardly observed and likely not to be politically desired, the penalty is assumed to be exogenous in the analysis. *Efficiency* characterizes the allocation function and audit probability for given penalty functions.

With respect to the *audit probability* it is the aim of the government on one hand, to keep the audits high in order to induce incentive compatibility, but on the other hand to keep the audits low in order to save audit costs. An efficient verification scheme therefore has the property, that, for a given penalty structure, incentive compatibility cannot be achieved with less auditing. If, in addition to the penalty function, the allocation function is given, the determination of the efficient audit probability is straightforward. Other things being equal, smaller audit probabilities are always preferred, as they produce lower audit costs. Total audit costs will be as low as possible, if they are as low as possible for every single firm.

With respect to the *allocation function* the aim is to maximize the allowances that are allocated to every single firm (provided the firm has reported truthfully). This target may be motivated by lobbying power of the firms or, in an international context, the wish to promote national firms. In the EU Emissions Trading Scheme, for example, it can be observed that the EU commission constantly has to reprove member countries for wishing to allocate too many allowances. The technical rationale of this condition is that for an allocation of allowances other than the maximum possible one, audit costs would be higher than necessary for obtaining incentive compatibility, which, of course, can not be efficient. For a given penalty function and audit strategy the structure of an efficient allocation function can thus be determined.

The formal definition of efficiency is given by the following:

For any given penalty function $f(\cdot)$, a mechanism $(e, a, f) \in F_I$ is *efficient* in F_I if there is no other mechanism $(e', a', f) \in F_I$ with

$$\begin{aligned} & e'(\cdot) \geq e(\cdot),\ a'(\cdot) \leq a(\cdot) \text{ and } a'(\cdot) \neq a(\cdot), \\ \text{or} \quad & e'(\cdot) \geq e(\cdot),\ e'(\cdot) \neq e(\cdot) \text{ and } a'(\cdot) \leq a(\cdot). \end{aligned} \tag{2.18}$$

A distinctive feature of the problem under study is the already mentioned cap on overall allowances. Incentive compatibility and efficiency determine the structure of an optimal verification scheme under the assumption that there is no cap on allowances. In order to take account of this constraint, it then has to be checked, which of the optimal schemes will meet the cap, i.e. whether the sum of allocated allowances equals the amount available (recall that the allocation function determines the amount of emission allowances to be allocated for each report of each firm). This result will depend, of course, on the underlying distribution of the emission values. In most of the following analysis the focus will therefore be on determining the structure of efficient functions and on the question how the government should design the verification scheme, such that incentive compatibility and efficiency are achieved.

This constitutes no restriction, since once the structure is known, an efficient scheme can be made to meet the cap without changing the incentives nor losing efficiency, if the allocation and the penalty function are multiplied with a respective factor, δ, that ensures that the allowances allocated equal the amount available: $\pi_\delta(\hat{\theta}, \theta) := \delta e(\hat{\theta}) - a(\hat{\theta})\delta f(\hat{\theta}, \theta) = \delta(e(\hat{\theta}) - a(\hat{\theta}) f(\hat{\theta}, \theta)) = \delta\pi(\hat{\theta}, \theta)$ and therefore $\pi_\delta(\theta, \theta) \geq \pi_\delta(\hat{\theta}, \theta) \Leftrightarrow \pi(\theta, \theta) \geq \pi(\hat{\theta}, \theta)$ for each $\delta \in \mathbb{R}_+$. Thus, as long as there are no political objections against a change in the size of the penalty (which is likely to be the case as long as the change is moderate), it is possible to make an efficient scheme meet the cap on allowances.

2.4 General Efficiency Conditions

Incentive compatibility (2.16) is equivalent to

$$a(\hat{\theta}) f(\hat{\theta}, \theta) \geq e(\hat{\theta}) - e(\theta) \quad \forall\, \hat{\theta}, \forall\, \theta. \tag{2.19}$$

In order to induce truth-telling by the firms the government has to choose the allocation function $e(\cdot)$, the penalty function $f(\cdot)$, and the audit probability $a(\cdot)$ such that for each firm the expected fine is not smaller than the "gross gain from misreporting", i.e. the gain the firm made if there were no penalties or if it were not audited. Since $a(\cdot) \leq 1$ this implies in particular that for all mechanisms element F_I it holds $f(\hat{\theta}, \theta) \geq e(\hat{\theta}) - e(\theta)$ for all $\hat{\theta} \in [0, \theta_u]$. Note that firms will not underreport, since $e(\cdot)$ is nondecreasing and $f(\hat{\theta}, \theta) = 0$ for $\hat{\theta} < \theta$. In the following the analysis therefore focuses on ensuring (2.16) for $\hat{\theta} > \theta$.

A firm which reports a value $\hat{\theta}$ may thus truly have any emissions value $\theta < \hat{\theta}$. With respect to the audit probability that guarantees incentive compatibility, (2.16) implies

$$a(\hat{\theta}) \geq \sup_{\theta < \hat{\theta}} \frac{e(\hat{\theta}) - e(\theta)}{f(\hat{\theta}, \theta)} \quad \forall\, \hat{\theta}. \tag{2.20}$$

For any given pair of allocation and penalty functions, the audit probability for a reported type $\hat{\theta}$ has to be large enough to outweigh the potential expected gain for any true type $\theta < \hat{\theta}$. (2.20) states that this is achieved by setting the audit probability for any report at least equal to the largest possible ratio of gross gain to penalty for an overstatement.

Efficient mechanisms can now be characterized:

Proposition 1 *1. The audit probability of an efficient mechanism* $(e, a, f) \in F_I$ *is given by*

$$a(\hat{\theta}) = \sup_{\theta < \hat{\theta}} \frac{e(\hat{\theta}) - e(\theta)}{f(\hat{\theta}, \theta)} \quad \forall\, \hat{\theta}. \tag{2.21}$$

2. Let $e(\cdot)$ *be any allocation function that is continuous and fulfills the feasibility requirements. If* $\arg\sup_{\theta<\hat{\theta}} \frac{e(\hat{\theta})-e(\theta)}{f(\hat{\theta},\theta)} < \hat{\theta}$ *in (2.21), the condition is also sufficient for efficiency.*

3. If $\arg\sup_{\theta<\hat{\theta}} \frac{e(\hat{\theta})-e(\theta)}{f(\hat{\theta},\theta)} = \hat{\theta}$*, sufficiency of (2.21) requires in addition differentiability of* $e(\hat{\theta})$ *and* $f(\hat{\theta}, \theta)$, $f_\theta(\hat{\theta}, \theta) < 0$ *and* $a(\hat{\theta}) \not\equiv const.$

The proof is given in Appendix B. The necessity of (2.21) is intuitive. As stated above, the audit probability has to be large enough to induce incentive compatibility for any possible type $\theta < \hat{\theta}$. However, an audit probability that is higher than this minimal necessary one is inefficiently high, in the sense that incentive compatibility can already be achieved with less auditing and therefore less audit costs. Efficiency can therefore only be achieved if equality holds in (2.20).

Note that there is no restriction on the form of the allocation function. In particular the function need not necessarily be concave, which is a major difference to results in for example the tax evasion literature, cf. section 3.5. The indeterminacy of the allocation function is explained by the "lack of conflict" in the efficiency concept, since the low audit probability as well as the maximization of allocations to firms that have reported truthfully are in the interests of both the government and the firms.

If $a(\hat{\theta}) = const$, efficiency will be determined by the allocation function $e(\cdot)$. Such a scenario will be analyzed in more detail in section 3.4 on Full Verification and in chapter 4.

Proposition 1 makes the strong interrelation apparent that the three functions have in an efficient verification scheme. Not surprisingly is the audit probability lower for a higher penalty. Fines should be substituted for audits, since audits are costly, and fines are not. As mentioned above, this reflects the well known result that enforcement costs are kept at the minimum, if penalties for misreporting are increased as far as possible and thus the probability of costly auditing is as small as possible (BECKER, 1968). In condition (2.8) boundedness of the penalty function is required, the schemes I consider are therefore necessarily second-best. Without other restrictions, increasing the penalties as far as possible would imply that they are equal to this upper bound for every overstatement. Such a result can be found for example in CHANDER AND WILDE (1998) in a tax evasion context, or in POLINSKY AND SHAVELL (2000) in a crime and punishment framework. In order to allow for a differentiation of the penalty with respect to the size of the overstatement, I introduce specific penalty structures.

2.5 Penalty Functions

Six specific penalty structures are chosen to point out main features of an efficient verification scheme. In this section these penalties are introduced and placed in the context of different strands of literature, namely tax evasion, crime and punishment, compliance and enforcement, and natural resource enforcement problems.

Proportional Penalties

For the major part of the analysis linear penalty functions, corresponding to a *penalty that is proportional to the overstatement*, will be used: For $\gamma \in \mathbb{R}_+$ they are given by:

$$f_{prop}(\hat{\theta}, \theta) = \max\{0, \gamma(\hat{\theta} - \theta)\}. \tag{2.22}$$

The idea behind such a penalty structure is that higher absolute deviations from the true emissions value should be penalized stronger, independent of whether the true emissions value is high or low.

Proportional penalties lead to expected allowances of

$$\pi_{prop}(\hat{\theta}, \theta) = \begin{cases} e(\hat{\theta}) - a(\hat{\theta})\gamma[\hat{\theta} - \theta] & \text{if } \hat{\theta} > \theta \\ e(\hat{\theta}) & \text{if } \hat{\theta} \leq \theta. \end{cases} \tag{2.23}$$

Penalties that depend linearly on the difference between the true and the reported value can be found for example in the tax evasion literature in the basic analysis by ALLINGHAM AND SANDMO (1972), and in CHANDER AND WILDE (1998).

Constant Penalty

The rationale behind the second penalty structure is the idea that rough estimates of the data may be obtained at low costs, whereas it will be very time and cost consuming to receive the exact values. In this case it may be desirable to allocate constant fines to everybody who is found to have overstated his emissions, since this would allow for rough estimates. For $C \in \mathbb{R}_+$ the *constant penalty* is given by:

$$f_{const}(\hat{\theta}, \theta) = \begin{cases} C & \text{if } \hat{\theta} > \theta \\ 0 & \text{if } \hat{\theta} \leq \theta \end{cases} \tag{2.24}$$

and leads to expected allowances of

$$\pi_{const}(\hat{\theta}, \theta) = \begin{cases} e(\hat{\theta}) - a(\hat{\theta})C & \text{if } \hat{\theta} > \theta \\ e(\hat{\theta}) & \text{if } \hat{\theta} \leq \theta. \end{cases} \tag{2.25}$$

In the tax evasion literature REINGANUM AND WILDE (1985) use constant taxes and fines for their comparison of audit cutoff versus random audit rules. DYE (1986) analyzes optimal monitoring policies in a general principal-agent framework with constant payments.

Zero Allowances

Note the difference between constant penalties and a penalty that leaves the firm with zero allowances in case of a detected overstatement:

$$f_{zero}(\hat{\theta},\theta) = \begin{cases} e(\hat{\theta}) & \text{if } \hat{\theta} > \theta \\ 0 & \text{if } \hat{\theta} \leq \theta \end{cases}, \tag{2.26}$$

Such a structure would result if penalties were determined endogenously under the condition that expected allowances don't become negative (since $e(\hat{\theta}) - a(\hat{\theta})f(\hat{\theta},\theta) \geq 0 \Rightarrow f(\hat{\theta},\theta) \leq e(\hat{\theta})$). In the tax evasion literature, it can be found in CHANDER AND WILDE (1998) or MOOKHERJEE AND PNG (1989). In this model, however, negative expected allowances are not ruled out, which leads to different results. I will comment on this structure and compare the results with corresponding results in the tax evasion literature in chapter 3.5.

Correct and Penalize

Penalties that depend solely on the degree of misreporting have one apparent drawback. The knowledge of the true emission value θ, that is obtained by the audit, is only used for the penalty function, not for the initial allocation. Preferable would be an "allocation after audit" of the form $e(\theta) - f(\hat{\theta},\theta)$ instead of $e(\hat{\theta}) - f(\hat{\theta},\theta)$, i.e. a correction for the gain of overstating. I therefore deviate from the model framework and analyze a *penalty that corrects and penalizes*

$$\tilde{f}(\hat{\theta},\theta) = \max\{0, (e(\hat{\theta}) - e(\theta) + f(\hat{\theta},\theta))\}, \tag{2.27}$$

where $f(\hat{\theta},\theta)$ is a penalty function of the form described above, for example a proportional or a constant penalty. Expected allowances are then given by:

$$\tilde{\pi}(\hat{\theta},\theta) = e(\hat{\theta}) - a(\hat{\theta})\tilde{f}(\hat{\theta},\theta) \tag{2.28}$$

$$= \begin{cases} (1-a(\hat{\theta}))e(\hat{\theta}) + a(\hat{\theta})(e(\theta) - f(\hat{\theta},\theta)) & \text{if } \hat{\theta} > \theta \\ e(\hat{\theta}) & \text{if } \hat{\theta} \leq \theta. \end{cases} \tag{2.29}$$

The "tilde"-notation indicates, that this penalty function constitutes a slightly different model framework. Whenever conditions (2.5) to (2.11) hold in the "standard" framework, they hold in the "tilde" framework as well. (2.5) to (2.9) are evident. (2.10) holds since $\tilde{\pi}(\hat{\theta},0) \leq 0$ is equivalent to $(1-a(\hat{\theta}))e(\hat{\theta}) \leq a(\hat{\theta})f(\hat{\theta},0)$. Using $(1-a(\hat{\theta}))e(\hat{\theta}) \leq e(\hat{\theta})$, the result follows from condition (2.10) in the "standard" framework: $\pi(\hat{\theta},0) \leq 0 \Leftrightarrow e(\hat{\theta}) \leq a(\hat{\theta})f(\hat{\theta},0)$. (2.11) holds since $\tilde{f}(x,z) \leq \tilde{f}(x,y) + \tilde{f}(y,z)$ for $x \geq y > z$ is equivalent to $e(x)-e(z)+f(x,z) \leq e(x)-e(y)+f(x,y)+e(y)-e(z)+f(y,z) \Leftrightarrow f(x,z) \leq f(x,y) + f(y,z)$ for $x \geq y > z$.

However, $e(\cdot)$ nondecreasing cannot be shown and has to be adopted as an assumption. Let

$$\tilde{F} = \{(e,a,\tilde{f}) : (2.5) - (2.11) \text{ and } e \text{ nondecreasing hold for } (e,a,\tilde{f}), \\ \text{where } \tilde{f}(\hat{\theta},\theta) = \max\{0, (e(\hat{\theta}) - e(\theta) + f(\hat{\theta},\theta))\}\}. \tag{2.30}$$

The revelation principle holds then for mechanisms $(e,a,\tilde{f}) \in \tilde{F}$, which allows the concentration of the analysis on incentive compatible mechanisms.

Penalties proportional to the gain from the overstatement

In the fields of optimal tax design and of forest use contracts respectively, YITZHAKI (1974) and AMACHER ET AL. (2005) find that results for penalties that are proportional to the amount understated crucially differ from the results for penalties that are proportional to the amount of tax/payments evaded. In this context, penalties that are proportional to the gain from an overstatement are given by:

$$\tilde{f}_{propgain}(\hat{\theta},\theta) = \max\{0, \gamma(e(\hat{\theta}) - e(\theta))\}, \tag{2.31}$$

for $\gamma \in \mathbb{R}_+$. Thus they are a special case of correcting penalties. Using Proposition 1 it is easily seen that they lead to a constant audit probability.

Remark 2 *Penalties that depend on the gain of an overstatement instead of the amount overstated imply that the efficient audit probability is the same for every report:*

$$a(\hat{\theta}) = \sup_{\theta<\hat{\theta}} \frac{e(\hat{\theta}) - e(\theta)}{\gamma(e(\hat{\theta}) - e(\theta))} = \frac{1}{\gamma} \quad \forall\, \hat{\theta}. \tag{2.32}$$

This is not the case for the other introduced penalties, which depend on the amount overstated, as is shown in the following chapter. In (2.32) it is easily seen that the actual penalty has to be at least as big as the gain from the overstatement in order for the mechanism to be incentive compatible, i.e. $\gamma \geq 1$ holds for the penalty parameter.

Convex Penalties

Convex penalties are well known in the literature on compliance and enforcement, as for example in ARGUEDAS AND HAMOUDI (2004), where a policy is composed of pollution standards, costly inspection processes and fines for exceeding the standard. In the context of the grandfathering of emission allowances, however, the revelation principle does not hold for convex penalties, which means that the restriction on incentive compatible mechanisms might place a significant constraint on the analyzed set of mechanisms. Note however, that in environmental regulation problems the focus is on social welfare maximization, and not on efficient auditing. Under certain restrictions optimality in that sense is even achieved without incentive compatibility. This is due to the fact that in the compliance and enforcement context the compliance with the standard imposes costs on firms, whereas in my context truth-telling imposes no costs on the firms compared with overstating.

This shows that the problem I consider should not be compared with compliance and enforcement, but rather with tax evasion and crime problems.

2.6 Rent Seeking with Implicit Costs

Since for every country there is a fixed amount of allowances that is distributed to the firms, the problem seems to be a rent-seeking problem: every firm maximizes its own share of the "rent". Yet, there is an important difference between the problem studied in this analysis and a rent-seeking problem, which is the nature of the costs that are devoted to obtaining this rent. To see the difference, recall the basic characteristics of the standard rent-seeking model. The research on the widely studied field of rent-seeking goes back to TULLOCK (1967) and KRUEGER (1974), who first introduced the term "rent-seeking". A common feature of these analyses, as well as the broad subsequent literature (see e.g. LOCKARD AND TULLOCK (2001) for a compendium), is that they show that whenever people interact strategically to compete for some rent, they use up resources that would not be spent otherwise, which yields a welfare loss. This waste of resources is generally referred to as dissipation. In his seminal work, TULLOCK (1980) assumed rent-seeking games to have an additive cost structure. Analyzing the strategic interaction between players, reaction functions are then found to be non-constant.

In the standard Tullock model of rent-seeking, (TULLOCK, 1980), n rent seekers (here firms) compete for the rent with value V. Expected profits for firm i would then be

$$\Pi^i = p^i V - x^i, \tag{2.33}$$

where p^i is firm i's probability of success and x^i are firm i's costs.

In the context considered in this analysis, $p^i V$ can be interpreted as firm i's share of overall allowances V. However, the "costs" of this model differ from the costs assumed in the standard rent-seeking model. In the Tullock model the costs are given by the effort that the firm exerts to obtain the rent. In the case of the grandfathering of emissions rights, it is the report of the firm that can be interpreted as the effort. Making a statement on emission values does, however, not impose any costs on the firm in this model, independent of whether the statement is true or false. The costs that the firm has to bear for announcing a too high emission value are represented by a cut of its allowances as given by the penalty function, and they only apply if the firm is audited. The withheld allowances are then redistributed among all participating firms. This marks an important difference to the standard rent-seeking model, since it induces that expected profits are given by

$$\Pi^i := \Pi(\hat{\theta}^i, \hat{\theta}^{-i}) = p(\hat{\theta}^i, \hat{\theta}^{-i})V, \tag{2.34}$$

where $\hat{\theta}^i$ is firm i's report/effort and $\hat{\theta}^{-i} = (\hat{\theta}^1, \ldots, \hat{\theta}^{i-1}, \hat{\theta}^{i+1}, \ldots, \hat{\theta}^n)$ are the other firms' reports.

Costs are no longer additive, but directly influence firm i's share $p^i := p(\hat{\theta}^i, \hat{\theta}^{-i})$ of overall allowances. In addition, this share is influenced by the other firms' reports. To see this, consider another way of writing it: each firm expects to receive a certain amount of allowances, given by (2.4). The share $p^i V$ is then given by the product of expected allowances and a compliance factor $\delta(\cdot)$ that ensures that the overall allowances V are met:

$$\Pi^i = p(\hat{\theta}^i, \hat{\theta}^{-i})V = \delta(\hat{\theta}^i, \hat{\theta}^{-i})\pi(\hat{\theta}^i, \theta^i), \tag{2.35}$$

$$\text{with } \delta(\hat{\theta}^i, \hat{\theta}^{-i}) \sum_{j=1}^{n} \pi(\hat{\theta}^j, \theta^j) = V. \tag{2.36}$$

Expected allowances of each firm are increasing in the firms' report and decreasing in the expected penalty. Maximization of profit

$$\Pi(\hat{\theta}^i, \hat{\theta}^{-i}) = \delta(\hat{\theta}^i, \hat{\theta}^{-i})\pi(\hat{\theta}, \theta^i) = V\frac{\pi(\hat{\theta}^i, \theta^i)}{\sum_j \pi(\hat{\theta}^j, \theta^j)} \tag{2.37}$$

yields[5]

$$\begin{aligned}\frac{\partial \Pi(\hat{\theta}^i, \hat{\theta}^{-i})}{\partial \hat{\theta}^i} &= V\frac{\frac{\partial \pi(\hat{\theta}^i, \theta^i)}{\partial \hat{\theta}^i}\sum_j \pi(\hat{\theta}^j, \theta^j) - \pi(\hat{\theta}^i, \theta^i)\frac{\partial \pi(\hat{\theta}^i, \theta^i)}{\partial \hat{\theta}^i}}{(\sum_j \pi(\hat{\theta}^j, \theta^j))^2} \\ &\stackrel{!}{=} 0,\end{aligned} \tag{2.38}$$

$$\Leftrightarrow \quad \frac{\partial \pi(\hat{\theta}^i, \theta^i)}{\partial \hat{\theta}^i}\left(\sum_j \pi(\hat{\theta}^j, \theta^j) - \pi(\hat{\theta}^i, \theta^i)\right) = 0 \tag{2.39}$$

$$\Leftrightarrow \quad \frac{\partial \pi(\hat{\theta}^i, \theta^i)}{\partial \hat{\theta}^i} = 0 \quad \text{or} \quad \sum_{j \neq i} \pi(\hat{\theta}^j, \theta^j) = 0 \quad \forall i, j. \tag{2.40}$$

However,

$$\sum_{j \neq i} \pi(\hat{\theta}^j, \theta^j) = \sum_{j \neq i} e(\hat{\theta}^j) - a(\hat{\theta}^j)f(\hat{\theta}^j, \theta^j) = 0 \quad \forall\, i \tag{2.41}$$

never holds, from the assumption that $e \not\equiv 0$ and since the penalty is zero for $\hat{\theta} = \theta$.

[5] For this exercise assume that all functions are differentiable. Note that this is not required in the efficiency analysis.

Therefore the profit maximization condition in the "rent-seeking with implicit costs" setting equals the expected allowances maximization condition in the principal-agent setting: firms do not take their rivals' actions into consideration, the reaction function of each player is constant. This is due to the fact that the penalty function, and therefore the expected allowances/payoff, does not depend on the other firms' reports. Following KRUEGER (1974), who considers "ways in which rent seeking is competitive", one may denote the current analysis as "non-competitive rent-seeking". The basic results are summarized in

Remark 3 *Despite of the fixed amount of allowances the problem is not a rent-seeking problem, the firms' reaction functions are constant. In determining their optimal report* $\hat{\theta}$*, firms base their behavior on their true parameter value as well as the allocation, audit and penalty functions which are given by the government. By construction, no dissipation occurs.*

2.7 Summary

The task of this chapter was twofold. Firstly, the model framework was set up, with which the question of an efficient verification of firm data can be answered for the grandfathering process of an emissions trading scheme. Secondly, efficient verification schemes were characterized, and, by setting up the model and establishing general efficiency results, differences to other strands of literature were identified.

The guidelines for the model setup stem from the political reality. The, as obvious as vital, cornerstones for the model are:

- firms have private information concerning their emissions data,
- emission allowances are grandfathered free of charge based on firm reports,
- auditing is costly.

These cornerstones already imply the assumption that firms have an exact knowledge of their emissions data, which might be considered critical from e.g. an engineering perspective. In the economic analysis, however, it allows to speak of emission values instead of emission intervals.

Throughout the chapter it was established, that a principal-agent model similar to those in a tax evasion context is most suitable. The analysis therefore uses verification schemes which allocate emission allowances on the basis of the reported emission values of the firms, set audit probabilities which depend as well on the reported values, and impose penalties for misreporting.

The guidelines are further supplemented by the following: Allowances will not be allocated for a report of zero emissions, but for reports of positive amounts of emissions. The penalty is nondecreasing in the degree of misreporting and bounded. There is no reward for underreporting or truth-telling.

With respect to the objectives of the single actors, it holds:

- the government aims at high verification standards and low audit costs,
- firms maximize expected emission allowances,
- the government also maximizes the allowances for every single firm that has reported truthfully.

High verification standards are in the model represented by incentive compatibility and an equivalent of the revelation principle was established to justify the concentration on incentive compatible mechanisms. Due to the requirements of no reward for underreporting or truth-telling and of no allowance assignments for a report of zero emissions, the standard arguments for the proof of the revelation principle do, just as expected from the literature, not hold. Two additional conditions are identified, which are necessary for the concentration of the analysis on incentive compatible mechanisms: the request of incentive compatibility for a firm of type zero, and a constraint for the penalty function that results in subadditivity for the functions used in this analysis.

It seems politically desired that the allocation function is increasing (more correctly: nondecreasing). For most of the analysis, however, this does not have to be introduced as an assumption into the model, since in the revelation principle equivalent it could be shown that this attribute holds for every allocation function of an incentive compatible mechanism.

Incentive compatible mechanisms have the property that the expected fine is not smaller than the gross gain from misreporting. In addition, since the allocation function is nondecreasing, firms have no incentive to understate their emissions.

Another very special feature of the analyzed problem is the cap on overall allowances. It was possible to show that although companies compete about a fixed amount of allowances, the problem is not a rent-seeking problem. Firms base their behavior on their emissions value as well as the functions of the verification mechanism, but not on the other firms' actions. In addition, neither the cap on allowances nor the distribution of emission values play a role for the characterization of incentive compatible or efficient verification schemes.

Finally, there are, of course, some assumptions, that help to keep the analysis tractable: all actors are risk neutral, the model is static, there is no commitment problem, and the true parameter is discovered when a firm is audited.

The analysis provides a full characterization of the audit probability and allocation function in efficient verification schemes. The audit probability equals the largest possible ratio of gross gain to penalty for an overstatement. In this generality, the result therefore replicates the Becker solution, proposing to substitute (socially) costless fines for costly audits. With respect to the allocation function, the word "characterization" does not seem to be very suitable, since the result is that there is no restriction on this function. However, this obviously implies the big advantage, that for any politically desired allocation of allowances an efficient verification scheme can be set up.

In the following chapters this general result will be further specialized for exogenously given penalties. Several penalty structures were therefore introduced: penalties that are increasing in the degree of misreporting and therefore hit large deviations stronger; constant penalties that hit small deviations stronger; penalties that leave firms with zero allowances; and penalties that correct for the gain of overstating. Finally, via convex penalties, the differences of this analysis and those in the field of compliance and enforcement were pointed out.

All in all the model setup and general results have disclosed several strands of literature that the analysis is only seemingly related to. A further comparison with the tax evasion literature will be provided in the following chapter.

2.8 Incentives to Overstate

In the following chapter the result will be established, that efficient auditing in situations where agents have incentives to understate, as for example in a tax evasion context, differs strongly from efficient auditing where agents have incentives to overstate their private data. This suggests that it might be interesting to look out for other fields with incentives to overstate, in particular since to date there doesn't seem to be much literature on this special field of misreporting.

On a theoretical level, the payroll accounting system for the medical profession in Germany seems to be a prominent example. It describes a situation in which, similar to the grandfathering process, agents receive payments based on their own reports. They receive more the more they report and thus have incentives to overstate their private data. Auditing is necessary, but costly. There is a fixed amount of money to be distributed, but the number of recipients (doctors) seems to be too high for rent-seeking being an appropriate framework. Naturally, doctors maximize expected payments, authorities want to pay them well, but also "fair", in the sense that payments are justified and not given for services that have not been performed. In other words: incentive compatibility is desired. Rewards for truthful reporting seem unlikely, but may not be excluded. Just as in the grandfathering, or also the tax evasion context, the authority could install verification schemes as described above, that consist of the actual payment (here $e(\cdot)$), random audits (here $a(\cdot)$), and a penalty for misreporting (here $f(\cdot)$). Feasibility conditions (2.5) to (2.10) from this analysis can be transferred into the medical context. Condition (2.11) is likely to depend on whether rewards for truthful reporting would be excluded, as in the grandfathering context, or used as an instrument to induce incentive compatibility, as in the tax evasion context.

In praxis, however, there seem to be two major differences which make the problem different to the one in the emissions trading context. Firstly, doctors have upper limits for their reports that apparently are too strict to allow for significant deviations. Theoretically they could make overstatements in the bills for the patients that are privately ensured. However, and here the second difference comes in, they are allegedly more likely to perform services that would perhaps not have been strictly necessary from a medical point of view, than to simply misreport the services they have performed. This, however, may not be possible to verify.

Chapter 3

Random Auditing

The general relationship between the allocation function, audit probability and penalty function in an efficient verification scheme, that was established in the previous chapter, is now analyzed further for the above introduced penalty structures. In addition I compare the efficient solution to the currently used procedure, and compare the results with related fields of literature.

3.1 Proportional Penalty

"Necessity" results

Given the proportional penalty

$$f_{prop}(\hat{\theta}, \theta) = \max\{0, \gamma(\hat{\theta} - \theta)\} \tag{3.1}$$

with $\gamma \in \mathbb{R}_+$, it follows from Proposition 1 that the audit probability of an efficient mechanism $(e, a, f_{prop}) \in F_I$ is given by

$$a(\hat{\theta}) = \sup_{\theta < \hat{\theta}} \frac{e(\hat{\theta}) - e(\theta)}{\hat{\theta} - \theta} \frac{1}{\gamma} \quad \forall\, \hat{\theta}. \tag{3.2}$$

Also from Proposition 1 it follows that there is no restriction concerning the structure of an allocation function in an efficient verification scheme. In order to further specify the audit probability, assumptions are made for the moment about the allocation function:

Proposition 2 *Let $(e, a, f_{prop}) \in F_I$ be an efficient mechanism in F_I.*

1. *For concave allocation functions $e(\cdot)$, the audit probability $a(\cdot)$ is nonincreasing in the reported parameter and given by*

$$a(\hat{\theta}) = \frac{e(\hat{\theta})}{\hat{\theta}\gamma}. \tag{3.3}$$

2. *For convex functions $e(\cdot)$, it is nondecreasing and given by*

$$a(\hat{\theta}) = \frac{1}{\gamma} D^{-} e(\hat{\theta}), \tag{3.4}$$

where D^{-} denotes the left-hand derivative.

3. *In particular for linear allocation functions $e(\cdot)$, the audit probability $a(\cdot)$ is constant. Let $e(\hat{\theta}) = \lambda\hat{\theta}$, $\lambda \in \mathbb{R}_{+}$, then*

$$a(\hat{\theta}) = \frac{\lambda}{\gamma}. \tag{3.5}$$

Proof: $a(\hat{\theta}) = \sup_{\theta<\hat{\theta}} \frac{e(\hat{\theta})-e(\theta)}{\hat{\theta}-\theta}$ simply gives the slope of the secant through the points $(\theta, e(\theta))$ and $(\hat{\theta}, e(\hat{\theta}))$.
For $e(\cdot)$ *concave* and $\hat{\theta} \in [0, \theta_u]$ the slope through $(\theta, e(\theta))$ and $(\hat{\theta}, e(\hat{\theta}))$ with $\theta < \hat{\theta}$, $\theta \in [0, \theta_u]$, is steepest if $\theta = 0$. The supremum is therefore attained at $\theta = 0$ and (3.3) holds due to $e(0) = 0$. $a(\hat{\theta}) = \frac{e(\hat{\theta})}{\hat{\theta}\gamma}$ is a multiple of average allocations. Since $e(\cdot)$ is concave the average allocations, and therefore $a(\cdot)$, are nonincreasing in $\hat{\theta}$.
If $e(\cdot)$ is *convex* the supremum is attained at $\theta = \hat{\theta}$. $a(\hat{\theta})$ is given by a multiple of the left-hand derivative, and of the derivative where it exists. $e(\cdot)$ is convex, therefore $D^{-}e(\cdot)$ is nondecreasing in the reported parameter and it holds that $a(\cdot)$ is nondecreasing.
The linear case follows immediately. ∎

Corollary 1 *In an efficient mechanism $(e, a, f_{prop}) \in F_I$, regressivity of the allocation function $e(\cdot)$ already induces an audit probability of $a(\hat{\theta}) = \frac{e(\hat{\theta})}{\hat{\theta}\gamma}$.*

The result is proven in Appendix C by showing that $\sup_{\theta<\hat{\theta}} \frac{e(\hat{\theta})-e(\theta)}{\hat{\theta}-\theta} = \frac{e(\hat{\theta})}{\hat{\theta}\gamma}$ for regressive allocation functions $e(\cdot)$. Proposition 2 and Corollary 1 present necessary conditions for a mechanism $(e, a, f_{prop}) \in F_I$ to be efficient in F_I. Below I will analyze, whether these conditions are also sufficient. First, however, I present interpretations and implications of the results. As expected, a stronger penalty (i.e. a higher γ) leads to a lower audit probability. The interplay between initial allocations and audit probabilities depends on the form of the allocation function.

Regressive allocation of allowances

A *concave allocation function* describes a regressive allocation of allowances. The at first sight most striking result is that in an efficient mechanism the audit probability is nonincreasing (strictly decreasing in the case of a strictly concave allocation function) in the reported parameter. This means that companies which present higher emission reports are less likely to be audited, which seems to contradict the aim of detecting large deviations. One has to consider, however, that the penalty and allocation function were chosen in a way that firms have incentives to announce their true values. This implies that a higher report does not stand for a bigger false report, but for a higher true value. Reporting a bigger false report than the true value is too costly for the firms, since additional gains from overstating are outweighed by larger expected penalties. Auditing higher reports with a higher probability would then mean that those reports were audited with a probability that is higher than necessary for achieving incentive compatibility, which, of course, can not be efficient.

Equation (3.3) shows that the audit probability, thought of as a decimal number, should be given by a multiple of average allocations, where the multiplier is determined by the strength of the penalty function. Consider exemplarily the case that $\gamma = 1$, i.e. the penalty function is given by

$$f_{prop}(\hat{\theta}, \theta) = \max\{0, (\hat{\theta} - \theta)\}. \tag{3.6}$$

In this case the audit probability should equal average allocations. If, for example a firm reports the emission value $\hat{\theta} = 1000$ and allowances are allocated such that $e(1000) = 900$, the respective firm should be audited with a probability of 90%.

Figure 3.1 below displays a concave and nondecreasing allocation function $e(\cdot)$, corresponding to a regressive allocation of allowances. γ is assumed to be 1, therefore the audit probability is equal to the average allocation of allowances, as shown in equation (3.3). In the figure it is given by the slope of the straight line through (0,0) and $(\hat{\theta}^i, e(\hat{\theta}^i))$. It can easily be seen that a higher reported value (e.g. $\hat{\theta}^2$ instead of $\hat{\theta}^1$) leads to a lower audit probability.

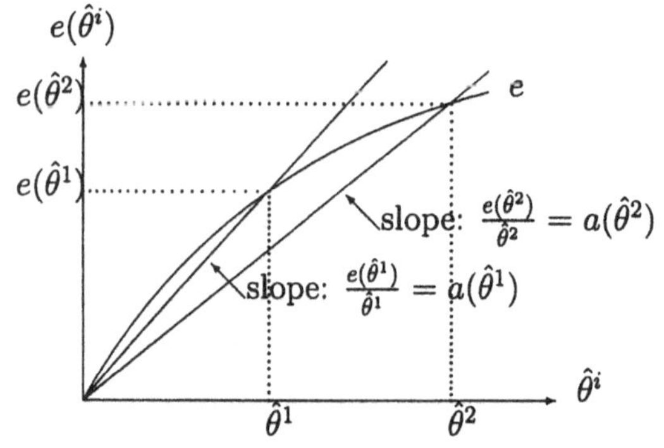

Figure 3.1: Regressive allocation of allowances

If the penalty is stronger than the difference between the reported and the true value, i.e. if $\gamma > 1$, the audit probability will be accordingly lower. Given the numerical example from above, a penalty factor of $\gamma = 10$ would reduce the audit probability that is necessary to induce incentive compatibility to 9%.

A simple insertion of a penalty factor $\gamma < 1$ seems to lead to contradictions in the above numerical example at first sight: a factor of $\gamma = 1/2$, for example, would lead to an audit probability of 180%, which is of course not possible. This example points up the significance of feasibility condition (2.10), which requires incentive compatibility for firms with no emissions ($\theta = 0$) and implies (2.12). For proportional penalties the latter is given by

$$e(\hat{\theta}) \leq a(\hat{\theta})\gamma\hat{\theta} \leq \gamma\hat{\theta} \quad \forall\, \hat{\theta} \in [0, \theta_u]. \tag{3.7}$$

It can immediately be seen that in the numerical example above $e(1000) = 900 \not\leq \frac{1}{2}1000 = 500$, thus such a mechanism is not feasible.

Condition (3.7) is helpful if it is read starting from the political guidelines. Suppose it had been decided that no firm (or e.g. no firm exceeding a certain firm size) should receive allowances for more than 50% of its emissions, i.e. $e(\hat{\theta}) \leq \frac{1}{2}\hat{\theta}$, and that fines should be proportional to the degree of misreporting, i.e. $f(\hat{\theta}, \theta) = f_{prop}(\hat{\theta}, \theta) = \max\{0, \gamma(\hat{\theta} - \theta)\}$. It then follows from above that γ has to be at least 1/2. Let, for example, $\gamma = \frac{1}{2}$. The audit probability should then be given by $a(\hat{\theta}) = e(\hat{\theta})/(\frac{1}{2}\hat{\theta})$ for each report $\hat{\theta}$, in order to achieve incentive compatibility and efficiency (strictly speaking, the latter only follows from the proof of sufficiency that is provided below in Proposition 4). Of course these considerations also hold for guidelines that fix the upper limit of allowances for a firm to a number that is above 100% of current emissions.

From the interrelation of initial allowances, penalty function and audit probability in efficient schemes as given by equation (3.3) it can be seen that the audit probability is increasing in the initial allowances, whereas a higher initial allocation of allowances has to be combined with higher penalties if audit probabilities shall not change.

Progressive allocation of allowances

A *convex allocation function*, on the other hand, describes a progressive allocation of emission allowances. Efficiency then implies the reverse effect on audit intensity: a plant which reports higher values will be audited with a higher probability. Equation (3.4) shows that the audit probability should be given by a multiple of the left-hand derivative, and of the derivative where it exists. The multiplier is again determined by the strength of the penalty function. The relationship between a convex allocation function and the audit probability in an efficient verification scheme is illustrated in Figure 3.2.

The displayed allocation function is differentiable, therefore the audit probability for report $\hat{\theta}^i$ is given by the first derivative of $e(\cdot)$ in $\hat{\theta}^i$ multiplied with $\frac{1}{\gamma}$. It is easily seen, that the audit probability is increasing in the reported parameter. Exemplarily it is shown to be higher for the report $\hat{\theta}^2$ than for the lower report $\hat{\theta}^1$.

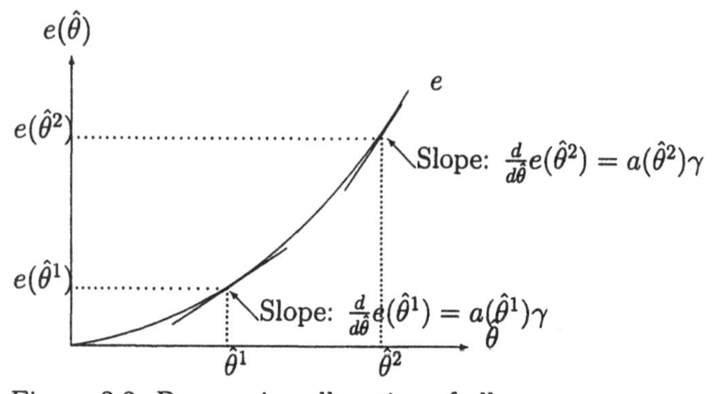

Figure 3.2: Progressive allocation of allowances

The incentive compatibility condition (2.16) can for convex allocation functions be reformulated to

$$a(\hat{\theta}) \geq \sup_{\theta<\hat{\theta}} \frac{e(\hat{\theta}) - e(\theta)}{\gamma(\hat{\theta} - \theta)} = \frac{1}{\gamma} D^{-}e(\hat{\theta}), \tag{3.8}$$

which is equivalent to

$$\gamma \geq D^{-}e(\hat{\theta}) \quad \forall\, \hat{\theta}. \tag{3.9}$$

This condition makes clear that only those mechanisms can be incentive compatible, which have an allocation function with a first (left-hand) derivative smaller than or equal to γ. The smaller the penalty the smaller the slope of the allocation function, and vice versa. Since all efficient mechanisms are in particular incentive compatible, the inequality naturally holds also for all efficient mechanisms. Again, starting with the political guidelines, (3.9) can be used to determine γ such that incentive compatibility holds.

Proportional allocation of allowances

The allocation schedule that is most popular in reality, is a proportional allocation of allowances. Such an allocation is described by a *linear allocation function.* Result (3.5) makes clear that the audit probability of an efficient mechanism with proportional allocations and penalties is the same for all reports. Furthermore, the audit probability is not equal to one, as in the schemes that can be observed in reality, but dependent on the allocation and penalty factor. If, for example, $e(\hat{\theta}) = 0.9\hat{\theta}$ for all $\hat{\theta}$ and $\gamma = 1$, the audit probability is 90% for all firms. Not surprisingly, it is increasing in the allocation factor and decreasing in the penalty factor. Figure 3.3 displays an exemplary allocation function $e(\hat{\theta}) = \lambda\hat{\theta}$ with $\lambda < 1$.

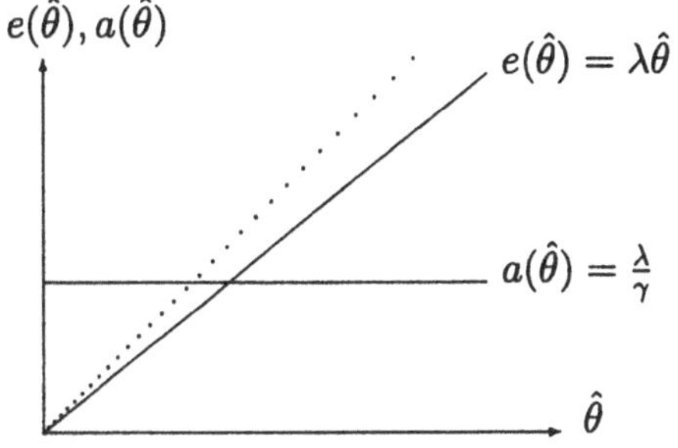

Figure 3.3: Proportional allocation of allowances

"Sufficiency" results

The proposition and corollary below show that (3.3) and (3.4) are also sufficient conditions for efficiency.

Proposition 3 *Let (e, a, f_{prop}) be a mechanism in F_I. It holds:*

1. *If*

$$a(\hat{\theta}) = \frac{e(\hat{\theta})}{\gamma\hat{\theta}} \quad \forall\, \hat{\theta} \in [0, \theta_u],$$

the allocation function $e(\cdot)$ is regressive.

2. *If*

$$a(\hat{\theta}) = \frac{1}{\gamma} D^{-} e(\hat{\theta}) \quad \forall\, \hat{\theta} \in [0, \theta_u],$$

the allocation function $e(\cdot)$ is convex, representing a progressive allocation of allowances.

The proof is given in Appendix C. The proposition shows that all verification schemes element F_I with an audit probability equal to a multiple of average (marginal) allocations, necessarily have a regressive (convex) allocation function. This is due to the incentive compatibility condition, which is not fulfilled otherwise. Given these results, it follows immediately:

Proposition 4 *Let* $(e, a, f_{prop}) \in F_I$ *and* $\gamma \in \mathbb{R}_+$. *Audit probabilities given by* $a(\hat{\theta}) = \frac{e(\hat{\theta})}{\gamma\hat{\theta}}$ *or* $a(\hat{\theta}) = \frac{1}{\gamma}D^- e(\hat{\theta})$ *are sufficient for a mechanism* $(e, a, f_{prop}) \in F_I$ *to be efficient in* F_I.

Proof: Let $a(\hat{\theta}) = \frac{e(\hat{\theta})}{\gamma\hat{\theta}}$. From Proposition 3 and $(e, a, f_{prop}) \in F_I$ it follows that $e(\cdot)$ is regressive. For continuous regressive functions $e(\cdot)$ with $e(0) = 0$ it holds that $\frac{1}{\gamma}\frac{e(\hat{\theta})}{\hat{\theta}} = \frac{1}{\gamma}\sup_{\theta<\hat{\theta}} \frac{e(\hat{\theta})-e(\theta)}{\hat{\theta}-\theta}$ for all $\theta, \hat{\theta} \in [0, \theta_u]$ (see the proof of Corollary 1 in Appendix C). Since $f_{prop} = \gamma(\hat{\theta} - \theta)$ for all $\hat{\theta} > \theta$, the result follows then immediately from Proposition 1.
Similar considerations hold for $a(\hat{\theta}) = \frac{1}{\gamma}D^- e(\hat{\theta})$. From Proposition 3 and $(e, a, f_{prop}) \in F_I$ it follows that $e(\cdot)$ is convex and therefore $a(\hat{\theta}) = \frac{1}{\gamma}D^- e(\hat{\theta}) = \frac{1}{\gamma}\sup_{\theta<\hat{\theta}} \frac{e(\hat{\theta})-e(\theta)}{\hat{\theta}-\theta}$ and the result follows from Proposition 1. ∎

An audit probability equal to a multiple of average or marginal allocations is sufficient to guarantee efficiency, if the penalty is differentiated with respect to the amount of misreporting. It cannot be excluded, however, that there are also other verification schemes that lead to efficiency, where the allocation function is not purely regressive or convex.

Efficiency in F

Since the revelation principle holds it is no loss of generality to focus the analysis on the class F_I of feasible and incentive compatible verification schemes. In order to illustrate if and in how far the assumptions that are necessary for the revelation principle to hold, i.e. (2.10) and (2.11), are restrictive, and whether efficiency conditions carry over from F to F_I, the analysis is at this point enlarged to the class F of feasible, but not necessarily incentive compatible mechanisms.

The penalty function is given exogenously, thus the only restrictive feasibility condition is (2.10), the condition that requires incentive compatibility for firms with zero emissions:

$$\pi(\hat{\theta}, 0) = e(\hat{\theta}) - a(\hat{\theta})\gamma\hat{\theta} \leq 0 \quad \forall\, \hat{\theta} \in [0, \theta_u].$$

Reformulating yields

$$a(\hat{\theta}) \geq \frac{e(\hat{\theta})}{\gamma\hat{\theta}} \tag{3.10}$$

for all $(e, a, f_{prop}) \in F$ and for all $\hat{\theta} \in [0, \theta_u]$.

For *concave allocation functions* it follows from (2.20) above that $a(\hat{\theta}) \geq \frac{e(\hat{\theta})}{\gamma\hat{\theta}}$ also has to hold for all $(e, a, f_{prop}) \in F_I$. This shows that incentive compatibility for the type $\theta = 0$, i.e. for firms with no emissions, already implies incentive compatibility for all types.

Remark 4 *Given a proportional penalty and a concave allocation function, condition (2.10) proves to be a strong condition; the sets F and F_I coincide.*

For *convex allocation functions* it follows from (2.20) that $a(\hat{\theta}) \geq \frac{1}{\gamma}D^{-}e(\hat{\theta})$ holds for all $(e, a, f_{prop}) \in F_I$. Since $D^{-}e(\hat{\theta}) \geq \frac{e(\hat{\theta})}{\hat{\theta}}$, this implies that F_I is a real subset of F. Figure 3.4 displays the affiliation of a verification scheme (e, a, f_{prop}) to F and F_I, depending on the audit probability $a(\cdot)$.

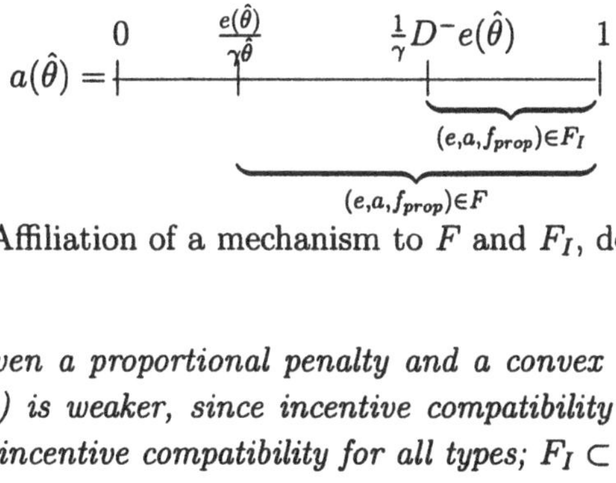

Figure 3.4: Affiliation of a mechanism to F and F_I, depending on $a(\cdot)$

Remark 5 *Given a proportional penalty and a convex allocation function, condition (2.10) is weaker, since incentive compatibility for the type $\theta = 0$ does not imply incentive compatibility for all types; $F_I \subset F$ holds.*

With respect to efficiency in F, the following statement can be shown:

Proposition 5 *A mechanism $(e, a, f_{prop}) \in F$ is efficient in F iff*

$$a(\hat{\theta}) = \frac{e(\hat{\theta})}{\gamma\hat{\theta}} \quad \forall\hat{\theta} \in [0, \theta_u]. \tag{3.11}$$

The result holds for any allocation function $e(\cdot)$ that fulfills the feasibility requirements.

The proof is straightforward and given in Appendix C. The comparison of mechanisms in F_I with respect to efficiency in F is given by the following

Corollary 2 *1. An efficient mechanism $(e, a, f_{prop}) \in F_I$ with $e(\cdot)$ concave and $a(\hat{\theta}) = \frac{e(\hat{\theta})}{\gamma\hat{\theta}}$ is also efficient in F, whereas a mechanism with e convex (not linear) and $a(\hat{\theta}) = \frac{1}{\gamma}D^{-}e(\hat{\theta})$ is efficient in F_I, but not in F.*

2. A mechanism $(e, a, f_{prop}) \in F_I$ is efficient in F only if $e(\cdot)$ is regressive.

Proof: 1. follows immediately, since $F_I \subseteq F$ and $D^{-}e(\hat{\theta}) \neq \frac{e(\hat{\theta})}{\hat{\theta}}$ for strictly convex functions. 2. follows from Proposition 5, which sets $a(\hat{\theta}) = \frac{e(\hat{\theta})}{\gamma\hat{\theta}}$, and Proposition 3, which states that in this case the allocation function is regressive for mechanisms element F_I. ■

An example

As an *example* consider the allocation function

$$e(\hat{\theta}) = \hat{\theta}^2$$

and audit probability

$$a(\hat{\theta}) = \frac{e(\hat{\theta})}{\gamma\hat{\theta}} = \frac{\hat{\theta}}{\gamma}.$$

From (2.12) the penalty parameter γ has to be chosen such that $e(\hat{\theta}) \leq \gamma\hat{\theta}$ for all $\hat{\theta} \in [0, \theta_u]$. For the chosen allocation function this implies that $\gamma \geq \theta_u$. It is evident, that the high penalty factor is due to the high progression. In reality one could imagine a range of sophisticated and/or piecewise progressive allocation function. For illustrative purposes, I stick to this simple form.

The verification scheme is element F by construction. To see that it is not incentive compatible, one can reason that (2.20) is not fulfilled, as $a(\hat{\theta}) = \frac{\hat{\theta}}{\gamma} < \frac{2\hat{\theta}}{\gamma} = \frac{1}{\gamma}D^{-}e(\hat{\theta})$, or, since all functions in the example are differentiable, calculate

$$\max_{\hat{\theta}} \pi(\hat{\theta}, \theta) = \hat{\theta}^2 - \frac{\hat{\theta}}{\gamma}\gamma(\hat{\theta} - \theta) = \hat{\theta}\theta. \tag{3.12}$$

For any given θ, the function $\pi(\hat{\theta}, \theta)$ is linear and increasing, therefore the maximum is attained at $\hat{\theta} = \theta_u$, and not at the true value $\hat{\theta} = \theta$, although for $\gamma = \theta_u$ this even implies an audit probability of one.

The given verification scheme is not element F_I, but the revelation principle states that there exists a mechanism $(e', a', f'_{prop}) \in F_I$ which replicates the equilibrium outcomes arising from (e, a, f_{prop}):

$$\pi'(\theta, \theta) = \pi(\alpha(\theta), \theta) \quad (3.13)$$

$$\text{and} \quad a'(\theta) = a(\alpha(\theta)). \quad (3.14)$$

Just like in the proof of Lemma 1 in Appendix B, denote the optimal report for the true type θ with $\alpha(\theta)$ and define

$$e'(\theta) := \pi(\alpha(\theta), \theta) = \theta_u \theta, \quad (3.15)$$

$$a'(\theta) := a(\alpha(\theta)), \quad (3.16)$$

$$f'(\hat{\theta}, \theta) := f(\hat{\theta}, \theta), \quad (3.17)$$

Note that the penalty function is the same in both schemes, in particular γ is the same. $a'(\theta) = a(\alpha(\theta))$ holds by definition, as well as $\pi'(\theta, \theta) = e'(\theta) = \pi(\alpha(\theta), \theta)$. The mechanism is incentive compatible, since $\pi'(\theta, \theta) = e'(\theta) = \theta_u \theta$ and $\pi'(\hat{\theta}, \theta) = \theta_u \hat{\theta} - \frac{\theta_u}{\gamma} \gamma(\hat{\theta} - \theta) = \theta_u \theta$ for all $\hat{\theta} \in [0, \theta_u]$.

From Proposition 5 it follows that $(e, a, f) \in F$ is efficient in F, since $a(\hat{\theta}) = \frac{e(\hat{\theta})}{\gamma \hat{\theta}}$ by construction. In addition, it follows from Proposition 2 that $(e', a', f') \in F_I$ is efficient in F_I, since $e'(\hat{\theta}) = \theta_u \hat{\theta}$ and $a'(\hat{\theta}) = \frac{\theta_u}{\gamma}$.

Summary

As argued above, high verification standards are in the model represented by incentive compatible verification schemes. The analysis has shown that under random auditing and proportional penalties there is no restriction with respect to the form of the allocation function, and allocation/audit-combinations that fulfill the required standards and in addition lead to efficiency under the above mentioned penalty include amongst others (i) a regressive allocation of allowances combined with an audit probability that is actually decreasing in the reported parameter and given by a multiple of average allocations, and (ii) a progressive allocation of allowances combined with an audit probability that is increasing in the reported parameter and given by a multiple of marginal allocations. This implies in particular, that for a proportional allocation of emission allowances, which can be observed in reality, the audit probability should be the same for all reports, i.e. there is no need to audit higher reports stronger, if the penalty and audit probability are chosen respectively.

If the analysis is extended to include the feasible but not incentive compatible verification schemes, it turns out that only those schemes can be efficient, which have an audit probability that equals a multiple of average allocations. The allocation function may, however, assume any form. Exemplary for a convex allocation function it was shown how the revelation principle can be applied to turn an efficient but not incentive compatible mechanism into one with the same outcomes that is incentive compatible and efficient among the incentive compatible mechanisms.

The results depend, of course, on the underlying penalty function. In the next section I therefore analyze a penalty function that does not differentiate with respect to the amount of misreporting.

3.2 Constant Penalty

In contrast to the proportional penalty, the constant penalty assigns the same penalty to every overstatement, independent of the degree of misreporting:

$$f_{const}(\hat{\theta}, \theta) = \begin{cases} C & \text{if } \hat{\theta} > \theta \\ 0 & \text{if } \hat{\theta} \leq \theta \end{cases}, \tag{3.18}$$

with $C \in \mathbb{R}_{+}$. Efficient mechanisms $(e, a, f_{const}) \in F_I$ can now be characterized as in the proportional penalty case:

Proposition 6 *A mechanism $(e, a, f_{const}) \in F_I$ is efficient in F_I iff*

$$a(\hat{\theta}) = \frac{e(\hat{\theta})}{C} \quad \forall\, \hat{\theta} \in [0, \theta_u]. \tag{3.19}$$

The audit probability is nondecreasing in the reported parameter. There is no restriction for the allocation function.

Proof: The results follow immediately from Proposition 1: $\sup_{\theta<\hat{\theta}} \frac{e(\hat{\theta})-e(\theta)}{C} = \frac{e(\hat{\theta})}{C}$, since $e(0) = 0$. The allocation function $e(\cdot)$ is nondecreasing and therefore $a(\cdot)$ is nondecreasing. ∎

Note that the expression for the audit probability does not depend on the form of the allocation function, as in the proportional penalty case. Firms which report higher values will be audited with a higher probability. This is explained by the fact that the penalty for an overstatement is independent of the deviation of the reported parameter from the true parameter. Therefore it is likely that firms which overstate their emissions will do so to a great extent. Thus the high reports should be audited with a higher probability. Although the fine as such is the same, the expected punishment for a smaller deviation would then be smaller than the expected punishment for a high deviation, due to the lower audit probability.

As expected, and just as for proportional penalties, the audit probability is negatively correlated to the strength of the penalty C. For high values of C, the audit probabilities will be low and vice versa. The parameter C is assumed to be given exogenously, but can be set by the government. It is evident that for constant penalties the penalty in case of an overstatement will be larger than the amount of allowances initially allocated. To see this, remember that $a(\hat{\theta}) \leq 1$ and therefore $C \geq e(\hat{\theta})$ for all $\hat{\theta} \in [0, \theta_u]$. This points out the necessity to deal with "negative allowances", as explained in chapter 2.

Note also that for any given C, every allocation function that fulfills the feasibility requirements (in particular in this case $e(\theta_u) \leq C$) can be used to set up an efficient mechanism by setting $f(\cdot) = f_{const}(\cdot)$ and $a(\hat{\theta}) = \frac{e(\hat{\theta})}{C}$. Only in very special cases will the audit probability be one. I will specialize these in chapter 3.4 on Full Verification.

It can easily be seen, that for constant penalties the sets F and F_I coincide. Since the penalty is the same for all overstatements, F_I requires that the audit probability hast to be high enough to guarantee incentive compatibility even for the biggest possible deviation. However, this is just the one for the type $\theta = 0$ and the condition for a mechanism to be element F.

3.3 Correct and Penalize

As explained in chapter 2.5, penalties that correct and penalize are a combination of a penalty structure analyzed up to now, and a second term that corrects for the gain that resulted if there were no audit:

$$\tilde{f}(\hat{\theta},\theta) = \max\{0, (e(\hat{\theta}) - e(\theta) + f(\hat{\theta},\theta))\}. \tag{3.20}$$

It constitutes a slightly changed model framework, but the general efficiency results that were obtained in chapter 2.4 carry over to this framework and efficient mechanisms $(e, a, \tilde{f}) \in \tilde{F}_I$ can be characterized as before:

Proposition 7 *The audit probability of an efficient mechanism* $(e, a, \tilde{f}) \in \tilde{F}_I$ *is given by*

$$a(\hat{\theta}) = \sup_{\theta<\hat{\theta}} \frac{e(\hat{\theta}) - e(\theta)}{e(\hat{\theta}) - e(\theta) + f(\hat{\theta},\theta)} \quad \forall\, \hat{\theta} \in [0, \theta_u]. \tag{3.21}$$

The result follows immediately from Proposition 1. Sufficiency of the condition also follows under the same restrictions as in Proposition 1. An audit probability as given by (3.21) is in this generality analytically less tractable and not interpretable. A closed term can be obtained if the penalty part $f(\hat{\theta},\theta)$ is further specified. For proportional or constant penalties results are similar to those obtained above:

Corollary 3 *Let* $(e, a, \tilde{f}) \in \tilde{F}_I$ *be an efficient mechanism in* $\tilde{F}_I$.

1. *For proportional penalties* $f_{prop}(\cdot)$ *and concave allocation functions* $e(\cdot)$, *the audit probability* $a(\cdot)$ *is nonincreasing in the reported parameter and given by*

$$a(\hat{\theta}) = \frac{e(\hat{\theta})}{e(\hat{\theta}) + \hat{\theta}\gamma}. \tag{3.22}$$

2. *For proportional penalties* $f_{prop}(\cdot)$ *and convex functions* $e(\cdot)$, *it is nondecreasing and given by*

$$a(\hat{\theta}) = \frac{e'(\hat{\theta})}{e'(\hat{\theta}) + \gamma}, \tag{3.23}$$

with $e'(\cdot)$ *the first derivative*[1].

[1] For ease of notation, the allocation function $e(\cdot)$ is assumed to be differentiable here.

3. *In particular for proportional penalties $f_{prop}(\cdot)$ and linear allocation functions $e(\cdot)$, the audit probability $a(\cdot)$ is constant. Let $e(\hat{\theta}) = \lambda\hat{\theta}$, $\lambda \in I\!R_+$, then*

$$a(\hat{\theta}) = \frac{\lambda}{\lambda + \gamma}. \tag{3.24}$$

4. *For constant penalties $f_{const}(\cdot)$ the audit probability $a(\cdot)$ is nondecreasing in the reported parameter and given by*

$$a(\hat{\theta}) = \frac{e(\hat{\theta})}{e(\hat{\theta}) + C}. \tag{3.25}$$

There is no restriction for the allocation function.

The results follow from plugging in the respective penalty function and considering where, with the respective allocation structure, the supremum is attained. Although the audit probability has naturally become smaller compared to the results with a penalty function without correction for gains of overstatements, the direction of change of the audit probability stays the same for each of the cases.

An interesting result can be obtained by comparing the penalty that corrects and penalizes to those penalties that depend solely on the degree of misreporting. Let $f(\hat{\theta}, \theta)$ be a penalty function with all properties that were established above. In particular let $f(\cdot)$ fulfill $a(\hat{\theta})f(\hat{\theta}, \theta) \geq e(\hat{\theta}) - e(\theta)$, i.e. incentive compatibility. It is obvious, that $\tilde{f}(\hat{\theta}, \theta) = \max\{0, (e(\hat{\theta}) - e(\theta) + f(\hat{\theta}, \theta))\}$ is stronger than $f(\hat{\theta}, \theta)$. The difference equals just the gain from misreporting, by construction. An estimate can then be given by how much this additional penalty can at least lower the audit probability that ensures incentive compatibility.

Proposition 8 *Let $(e, a, \tilde{f}) \in \tilde{F}_I$ be an efficient mechanism in $\tilde{F}_I$ and the associated $f(\cdot)$ strong enough to induce incentive compatibility. The audit probability is then smaller than 1/2.*

Proof: Incentive compatibility implies $f(\hat{\theta}, \theta) \geq e(\hat{\theta}) - e(\theta)$. Plugging into (3.21) leads to $a(\cdot) \leq 1/2$. ∎

As an example consider the proportional allocation $e(\hat{\theta}) = 0.9\hat{\theta}$ where each firm will receive allowances for 90% of its emissions. If the penalty were chosen to be a proportional penalty and the penalty factor $\gamma = 1$, the above analysis has shown that each firm should be audited with a probability of 90% in order to achieve incentive compatibility and efficiency. Under a penalty that corrects and penalizes, however, the efficient audit probability will be below 50%, as shown in Proposition 8. For concrete allocation and penalty functions it can be calculated and in the example it will be $a(\hat{\theta}) = \frac{0.9(\hat{\theta}-\theta)}{1.9(\hat{\theta}-\theta)} \approx 47\%$.

This result points up that the audit probability, and therefore the audit costs, can be decreased substantially when the knowledge of the true emissions value, that is obtained in the verification process, is not only used for the penalty, but also for the initial allocation. It also indicates that, although the audit probability of an efficient verification scheme with penalties that correct and penalize lacks the straightforward interpretation and thus seems to be difficult to communicate, such a scheme may still be preferable, due to the (possibly high) savings in audit costs. Of course a decrease in the audit probability can also be achieved by simply making the penalty factor γ bigger. One may imagine, however, that there might be more resentments against a "just very high" penalty than against a penalty that corrects for "undeserved gains".

3.4 Full Verification

From what can be observed during the initial stage of the EU emissions trading scheme, "high verification standards" were in most countries realized by an audit probability of 100% for every report, which I will call *Full Verification.* The analysis above has already shown, that audit costs can be reduced when the desired high verification standard is realized by achieving incentive compatibility via a scheme containing audits and penalties. Only under certain conditions for the allocation and penalty functions can Full Verification lead to efficiency. These are provided in this section.

It is shown that the form of the penalty determines the form of the allocation function, which in turn implies that the size of the penalty depends on the overall amount of allowances available. It is shown in Appendix C that:

Proposition 9 *1. A mechanism $(e, a, f_{prop}) \in F_I$ with $a(\cdot) \equiv 1$ is only efficient in F_I, if $e(\hat{\theta}) = \gamma\hat{\theta}$ for all $\hat{\theta} \in [0, \theta_u]$, where γ is the penalty parameter.*

2. A mechanism $(e, a, f_{const}) \in F_I$ with $a(\cdot) \equiv 1$ is only efficient in F_I, if $e(\cdot)$ is discontinuous with $e(0) = 0$ and $e(\hat{\theta}) = C$ for all $\hat{\theta} \in (0, \theta_u]$, where C is the value of the constant penalty.

Thus, for each given penalty structure, there is only one allocation function for which Full Verification is efficient in F_I. The Full Verification case therefore provides an example, where two of the three functions are given ($a(\cdot) \equiv 1$ and $f(\cdot) = f_{prop}(\cdot)$ or $f(\cdot) = f_{const}(\cdot)$) and, not surprisingly, uniquely determine the third function. Thereby the allocations depend on the report, when the penalty does, and they are constant for each report when the penalty is the same for all overstatements. The size of the penalty is important in so far as it directly influences the allowance allocation. If the upper limit of allowances is taken into account, the penalty parameters will have to be chosen such that this upper limit is not exceeded, i.e. such that $\sum_i e(\theta^i) = V$. This will of course depend on the underlying density function.

More generally it can be observed from Proposition 9 that under Full Verification and proportional penalties, the penalty equals for each type the gain from misreporting while for the discontinuous constant penalties it equals the maximum gain, i.e. the gain for type zero. Similar results are obtained for penalties that correct and penalize. The analysis for this penalty is slightly different, since it depends on the allocation function in addition to the penalty part.

Proposition 10 *1. A mechanism $(e, a, \tilde{f}_{corr}) \in \tilde{F}_I$ with $a(\cdot) \equiv 1$ and $f(\cdot)$ inducing incentive compatibility fails to be efficient in $\tilde{F}_I$.*

2. For a mechanism $(e, a, \tilde{f}_{corr}) \in \tilde{F}_I$ with $a(\cdot) \equiv 1$ to be efficient in $\tilde{F}_I$, there is no restriction concerning the allocation function. However, the mechanism is efficient if $f(\hat{\theta}, \theta) = 0$ for at least one $\theta \in [0, \hat{\theta})$ for every $\hat{\theta} \in [0, \theta_u]$.

The first result follows immediately from Proposition 8, which states that the given conditions imply an efficient audit probability not larger than 1/2.

The second statement follows from Proposition 7: $a(\hat{\theta}) = \sup_{\theta<\hat{\theta}} \frac{e(\hat{\theta})-e(\theta)}{e(\hat{\theta})-e(\theta)+f(\hat{\theta},\theta)}$ $= 1$ for all $\hat{\theta} \in [0, \theta_u]$. Thus, for every $\hat{\theta} \in [0, \theta_u]$, $f(\hat{\theta}, \theta) = 0$ for *one* $\theta \in [0, \hat{\theta})$ is sufficient for $a \equiv 1$ being efficient. However, it then saves debates about the adequate form of the penalty function if the "real" penalty part is set equal to zero for all reports. Thus, in order for an audit probability of one to be incentive compatible and efficient, the penalty may be set equal to the gain from the overstatement. Since $\pi(\hat{\theta}, \theta) = e(\hat{\theta}) - a(\hat{\theta})(e(\hat{\theta}) - e(\theta)) = e(\theta)$ for $a(\hat{\theta}) = 1$, the effect of combining Full Verification with such a penalty structure is the same as combining Full Verification with the credible statement that every firm will receive $e(\theta)$ after the audit. This shows two things: firstly, the system that was applied throughout the European Union at the start of the EU Emissions Trading Scheme is not inefficient with respect to the efficiency notion of this analysis. Secondly, it is quite obvious that still audit costs may be saved by introducing penalties that exceed the pure gain from the overstatement. This is no contradiction, since the focus of this analysis is on the determination of efficient schemes *given* a penalty structure. As noted above, if the aim of the government were a mere minimization of audit costs, the Becker-solution would naturally result.

3.5 Comparison with Tax Evasion

The reporting and verification of firm data in the grandfathering process of an emissions trading scheme displays strong parallels to the tax evasion problem. Not surprisingly many of the results in the two fields have similar structures. However, they also reveal important differences.

Chander and Wilde (1998), for example, analyze the interplay between optimal tax rates, audit probabilities and penalties for misreporting in a principal-agent model with a structure similar to the one of this analysis. In their notation the "penalty" is really a "post-audit payment", i.e. the payment after the audit has occurred and after the true income has been discovered. Let y denote income, x reported income, $t(\cdot)$ the tax function and $f_t(\cdot)$ the penalty function.

Proportional penalty

Under the penalty/post-audit payment

$$f_t(x,y) = t(x) + \max\{0, y - x\} \text{ for all } x \text{ and } y \tag{3.26}$$

the taxpayer will have to pay the tax $t(x)$ and all of the underreported income if he is audited. This penalty structure is comparable to the proportional penalty $f_{prop}(\hat{\theta}, \theta) = \max\{0, \gamma(\hat{\theta} - \theta)\}$ in my model, since this leads to a post-audit allocation with the same structure: $e(\hat{\theta}) + \max\{0, \gamma(\theta - \hat{\theta})\}$, due to

$$\begin{aligned} \pi_{prop}(\hat{\theta}, \theta) &= e(\hat{\theta}) - a(\hat{\theta}) \max\{0, \gamma(\hat{\theta} - \theta)\} & (3.27) \\ &= (1 - a(\hat{\theta}))e(\hat{\theta}) + a(\hat{\theta})(e(\hat{\theta}) - \max\{0, \gamma(\hat{\theta} - \theta)\}). & (3.28) \end{aligned}$$

The difference is in the interpretation: (3.26) determines a value that firms have to pay, whereas the parallel in the emissions trading model determines a value that firms receive. In the following I use the notation $x, y, t(\cdot)$ to indicate the tax evasion model, and $\hat{\theta}, \theta, e(\cdot)$ to refer to the emissions trading model. Chander and Wilde (1998) find, that efficiency is given only if the tax function $t(\cdot)$ is concave and nondecreasing and if the audit probability is nonincreasing. In particular the audit probability is determined by the marginal tax rate. These results differ from the results in the emissions trading model in two aspects.

Remark 6 *The audit probability $p(x)$ of an efficient tax/audit-scheme with a concave tax function is given by*

$$p(x) = D^+ t(x), \tag{3.29}$$

the right-hand derivative of the tax function, whereas the audit probability of an efficient allocation/audit-scheme with a concave allocation function is given by

$$a(\hat{\theta}) = \frac{e(\hat{\theta})}{\gamma \hat{\theta}}, \tag{3.30}$$

a multiple of average and not marginal allocations.

This difference is due to the fact that in the tax evasion context firms have an incentive to *understate* their income, whereas in the grandfathering context they have an incentive to *overstate* their emissions. Technically this leads to

$$p(x) = \sup_{y>x} \frac{t(y) - t(x)}{y - x} \tag{3.31}$$

versus

$$a(\hat{\theta}) = \sup_{\theta<\hat{\theta}} \frac{e(\hat{\theta}) - e(\theta)}{\gamma(\hat{\theta} - \theta)}. \tag{3.32}$$

At first sight, the efficient audit probabilities seem to be determined similarly: the fraction is in both cases the gain from the false report divided by the penalty. In the tax evasion case, however, the supremum is taken over all $y > x$. For every report x the government knows that the true value y is likely to be anything larger than the observed reported income. This is different in the grandfathering model: for every report $\hat{\theta}$ the government knows that the true emissions value θ is likely to be anything smaller than the observed one and thus the supremum over $\theta < \hat{\theta}$, which then causes the different results.

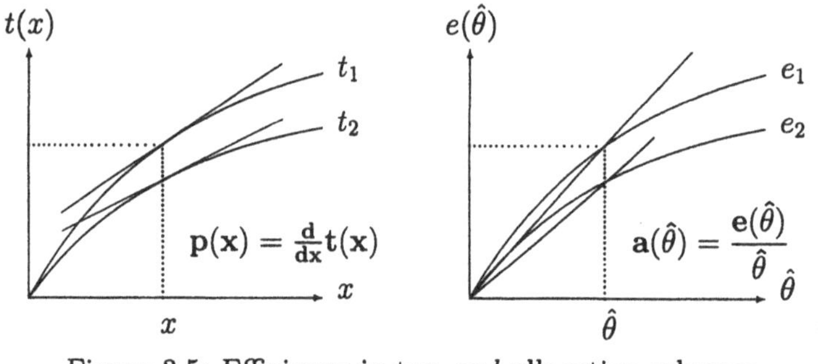

Figure 3.5: Efficiency in tax *and* allocation schemes

Figure 3.5 illustrates the results of Remark 6 with $\gamma = 1$. The graph on the right replicates the result established in Proposition 2 (see also equation (3.30) and (3.32)). The graph on the left demonstrates the respective result in the tax evasion model, which is given by equation (3.29) and (3.31). In both cases efficiency is achieved by setting the audit probability as indicated. The graphs also illustrate that, in both scenarios, the audit probability for a certain report increases, when the tax payment, respectively allowances allocation, for that report increases. In addition, for an otherwise similar setup, audit probabilities have to be higher when there is an incentive to overstate the private data. This holds since for a concave function the average is greater than the marginal value.

Another major difference is given by the number of efficient solutions:

Remark 7 *In the tax evasion context a mechanism is efficient only if* $t(\cdot)$ *is concave, whereas in the grandfathering context there is no such restriction for the allocation function. The equity-efficiency tradeoff that is present in the tax evasion context does not prevail in the grandfathering context.*

This results from the differences in the efficiency notion and also the direction of misreporting. Figure 3.6 presents an intuition for the different effect of the direction of misreporting on efficiency.

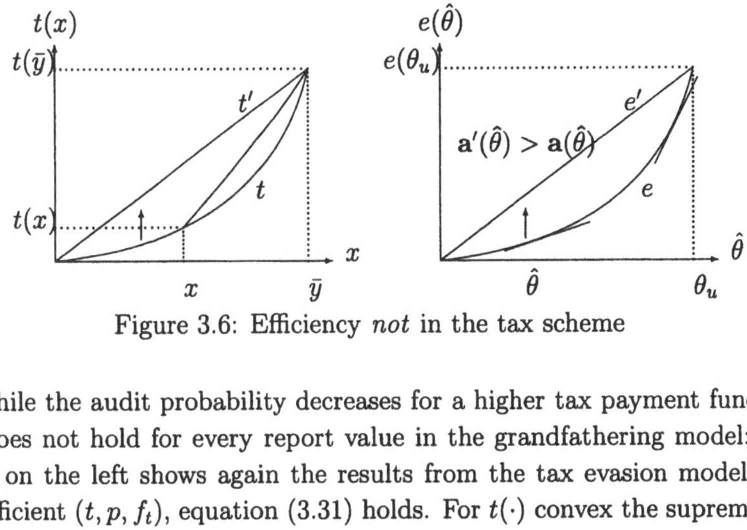

Figure 3.6: Efficiency *not* in the tax scheme

While the audit probability decreases for a higher tax payment function, this does not hold for every report value in the grandfathering model: The graph on the left shows again the results from the tax evasion model. For any efficient (t, p, f_t), equation (3.31) holds. For $t(\cdot)$ convex the supremum is attained at $y = \bar{y}$ and thus

$$p(x) = \frac{t(\bar{y}) - t(x)}{\bar{y} - x}. \tag{3.33}$$

Now a linear function $t'(\cdot)$ through the points $(0, t(0))$ and $(\bar{y}, t(\bar{y}))$ can be constructed. Being a linear combination of two points of a convex function, $t'(\cdot)$ is bigger than $t(\cdot)$, with $t'(\cdot) > t(\cdot)$ for $t(\cdot)$ strictly convex. With regard to the audit probability it holds

$$p'(x) = \frac{t(\bar{y})}{\bar{y}} < \frac{t(\bar{y}) - t(x)}{\bar{y} - x} = p(x), \tag{3.34}$$

for $t(\cdot)$ strictly convex. Thus there is a mechanism (t', p', f_t) with $t'(\cdot) > t(\cdot)$ and $p'(\cdot) < p(\cdot)$ for some x, i.e. there is no efficient (t, p, f_t) with $t(\cdot)$ strictly convex[2].

The graph on the right of Figure 3.6 replicates the results from Proposition 2. For any efficient $(e, a, f_{prop}) \in F_I$ equation (3.32) holds, the supremum is attained at $\theta = \hat{\theta}$ with $a(\hat{\theta}) = D^{-}e(\hat{\theta})$ and no function $e'(\cdot)$ can be found with $e'(\cdot) \geq e(\cdot)$, $a'(\cdot) \leq a(\cdot)$ and $e'(\cdot) \neq e(\cdot)$ or $a'(\cdot) \neq a(\cdot)$, since the audit probability is not smaller for every report $\hat{\theta} \in [0, \theta_u]$.

Excursus: Minimization of e(·)

At first sight the argument that prevented efficiency in the tax evasion model for a progressive tax scheme seems to be applicable to a regressive allocation of allowances in the grandfathering model. This "mirroring", however, is destroyed by the fact that in *both schemes* it is the aim of the government to *maximize* the functions $t(\cdot)$ and $e(\cdot)$ respectively. Just to see that even in the other case the mirroring of the models is not perfect and results are different, consider for the moment that it were the aim of the government to *minimize* the allocation function $e(\cdot)$. Figure 3.7 shows that then, indeed, a lower allocation function leads to a lower audit probability.

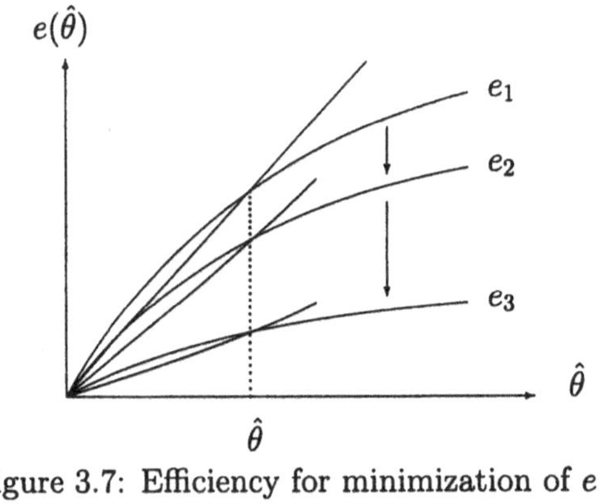

Figure 3.7: Efficiency for minimization of $e(\cdot)$

However, since reducing $e(\cdot)$ further also reduces the audit probability $a(\cdot)$ further, the only efficient mechanism would imply an allocation function equal (or since $e(\cdot) \not\equiv 0$ close) to zero. - End of minimization excursus.

[2] A formal proof of this result is provided in CHANDER AND WILDE (1998).

Remark 8 *The point that destroys the mirroring of arguments from the tax evasion to the grandfathering model is the fact that in both settings the audit probability is desired to be as low as possible in order to save audit costs.*

Penalty zero

The endogenous determination of the penalty function in CHANDER AND WILDE (1998) results in:

$$f_t(x, y) = y \text{ for } x \neq y, \tag{3.35}$$

i.e. if the firm is found to have misreported it will have to pay all of its income. In the emissions trading context such a penalty would be comparable to a penalty that takes away all of the allowances that the firm would have received otherwise and leaves it with zero allowances:

$$f_{zero}(\hat{\theta}, \theta) = \begin{cases} e(\hat{\theta}) & \text{if } \hat{\theta} > \theta \\ 0 & \text{if } \hat{\theta} \leq \theta \end{cases}, \tag{3.36}$$

leading to post-audit allowances of zero in case of an overstatement, and of $e(\hat{\theta})$ in case of $\hat{\theta} \leq \theta$. It is shown that

Remark 9 *Also for a penalty that leaves the firms with no income/allowances, the different direction of misreporting leads to different efficient audit probabilities. In particular they are equal to one for every report in the grandfathering model, but not in the tax evasion model.*

Incentive compatibility requires $e(\theta) \geq e(\hat{\theta}) - a(\hat{\theta})e(\hat{\theta})$ for all $\hat{\theta} > \theta$, and efficiency implies (cf. Proposition 1) that

$$a(\hat{\theta}) = \sup_{\theta < \hat{\theta}}(1 - \frac{e(\theta)}{e(\hat{\theta})}) = 1 \quad \forall\, \hat{\theta}, \tag{3.37}$$

because the supremum is attained at $\theta = 0$ for all $\hat{\theta}$. Although the penalty seems to be very strict, it has to be combined with a 100% audit probability in order to achieve incentive compatibility. It can easily be seen that this is due to firms with a small θ-value. For these firms the penalty is after all not very strong and if the audit probability were low it would be worthwhile to overstate to a great extent. This result is different from the result in the tax evasion model, where CHANDER AND WILDE (1998) find that the marginal payment rates determine the audit probabilities and thus efficient schemes may involve random or deterministic auditing.

Even if one argues that the endogenous penalty $f_t(\cdot)$ of the tax evasion model should be compared to the endogenously determined penalty of the grandfathering model, results concerning the efficient audit probability do not replicate. As argued in chapter 2, an endogenization of penalties would lead to a constant penalty with a penalty parameter C equal to the upper bound for every overstatement. As shown in chapter 3.2, random audit probabilities are not determined by marginal payment rates and they are deterministic only if $e(\hat{\theta}) = 0$ or $e(\hat{\theta}) = C$.

3.6 Summary

In this chapter efficient verification schemes were characterized for different penalty structures, resulting in diversified policy recommendations. The impact of the requirement of incentive compatibility has been studied, as well as that of a definite audit of every report. While in the previous chapter it was established that the proper model for this analysis is similar to those in a tax evasion context, in this chapter the main differences in the results were pointed out, as well as the critical assumptions causing them.

The influence of the penalty structure on the efficient audit probabilities is strong. While under penalties which are proportional to and increasing in the degree of misreporting, and therefore hit large deviations stronger, the form of the allocation function (i.e. progressive vs. regressive) is crucial for the question whether the audit probability should be decreasing or increasing in the reported value, this plays no role under constant penalties, where the audit probability is always nondecreasing in the reported value. For a proportional penalty and regressive allocation of initial allowances, for example, the efficient audit probability is shown to be nonincreasing in the reported emissions value and it is determined by average allocations, whereas the audit probability is the same for all reports, if allowances are allocated proportionally. In case of a progressive allocation of initial allowances the efficient audit probability is nondecreasing in the reported emissions value and determined by marginal allocations. Under constant penalties, which hit small deviations stronger, the audit probability depends positively on the allocation function and negatively on the value of the penalty.

Under both penalty structures the audit probability is, as expected, lower, the higher the penalty. The audit costs can therefore substantially be reduced when the knowledge of the true emissions value, that is obtained in the audit, is not only used for the penalty, but also for the allocation, in a way that results in a correction for the gain of overstating. The audit probability for a penalty that corrects and penalizes seems to be less calculable (and therefore likely to be more difficult to communicate to politicians or firms) at first sight. It provides, however, a simple structure for reducing audit costs. In addition the enlargement of the penalty in this special form can be motivated to "take away undeserved gains" and seems therefore likely to be highly acceptable.

While under constant penalties one of the conditions that had to be added to make use of the revelation principle enforces that only incentive compatible mechanisms are feasible, it is under proportional penalties also the form of the allocation function that determines, whether the requirement of incentive compatibility for firms of type zero puts a strong additional constraint on the set of feasible mechanisms, or not. If the allocation function is concave, the same restriction as for constant penalties applies. The condition is, however, less restrictive when the allocation function is convex. A necessary and sufficient condition for efficiency in F under proportional penalties is an audit probability determined by average allocations. This implies, that an incentive compatible mechanism can only be efficient among the non-incentive compatible mechanisms, if it contains a regressive allocation function.

All results do not depend on the underlying density function.

In addition, these proposed verification mechanisms have been compared to Full Verification, where every firm is definitely audited. Such a scenario allows to choose the allocation function freely, when the penalty is set equal to the gain from misreporting. If, on the other hand a certain penalty is fixed, it is the form of the penalty that determines the form of the allocation function.

Finally it was shown that although the structure of the model used for this analysis parallels the structure of principal-agent models in the tax evasion context, the results differ with respect to the number of efficient allocation (respectively tax) and audit - combinations, as well as the form of each variety. These differences are driven by two main factors: Firstly, in one case firms have an incentive to overstate, and in the other case to understate their private data, which influences the different forms. Secondly, the interests of the government and the firms, that result in the efficiency definition, differ in one case, and coincide in the other, which affects the number of efficient combinations. Another difference lies in the fact that rebates after an audit are by assumption excluded in the current model, while in the tax evasion literature several studies establish the result that an optimal verification scheme requires rebates for truth-telling.

Chapter 4

Deterministic Auditing

In practice, reports indicating high emissions seem to be audited very strictly, whereas low reports are treated rather laxly. In order to capture such an audit behavior I analyze verification schemes in which audit rules are different for emission reports above and below a given cutoff level. The previous chapter has shown, that it may not be efficient to audit high reports stronger. The question is therefore, how the verification scheme should be set up, such that the desired audit structure does not constitute a waste of audit resources, and under which conditions this deterministic audit structure should be preferred to random audits in terms of total audit costs.

4.1 Efficiency under different penalties

In verification schemes with *deterministic auditing* firms trigger an audit by reporting values above a given threshold. The audit probability is given by:

$$a_l(\hat{\theta}) = \begin{cases} 1 & \text{if } \hat{\theta} > l \\ 0 & \text{if } \hat{\theta} \leq l \end{cases} \tag{4.1}$$

Firms which report emission values above the cutoff level $l \in [0, \theta_u]$ will be audited with probability one, firms which report values below l will not be audited.

Expected allowances are then given by

$$\pi(\hat{\theta}, \theta) = e(\hat{\theta}) - a_l(\hat{\theta}) f(\hat{\theta}, \theta) = \begin{cases} e(\hat{\theta}) - f(\hat{\theta}, \theta) & \text{if } \hat{\theta} > l \\ e(\hat{\theta}) & \text{if } \hat{\theta} \leq l. \end{cases} \tag{4.2}$$

The only feasibility condition that places a restriction on the audit probability is condition (2.10), which requires that $\pi(\hat{\theta},0) = e(\hat{\theta}) - a_l(\hat{\theta})f(\hat{\theta},0) \leq 0$. The next aim is to construct the allocation function in a way that leads to incentive compatibility of the mechanism. Considering the fact that the true emissions value θ is not known, and therefore may be zero, and that $e(\cdot)$ is nondecreasing, this leads to

$$e(\hat{\theta})\begin{cases} \leq & f(\hat{\theta},0) \quad \text{if } \hat{\theta} > l \\ = & 0 \quad\quad\quad \text{if } \hat{\theta} \leq l \end{cases} \quad \text{for } (e, a_l, f) \in F. \tag{4.3}$$

This condition carries over the "cut" in the audit probabilities to the allocation function, since it implies that everybody who reports emission values below the cutoff level will not receive any allowances, since $\pi(\hat{\theta},\theta) = e(\hat{\theta}) = 0$, whereas everybody who does receive allowances will be audited with a probability of one. In addition the penalty for a firm of type zero has to be larger than the respective allocation, for every possible report.

The incentive compatibility condition $\pi(\theta,\theta) = e(\theta) \geq e(\hat{\theta}) - a_l(\hat{\theta})f(\hat{\theta},\theta) = \pi(\hat{\theta},\theta)$ can be set up just as in the random auditing case. Again, firms have no incentive to understate, since the allocation function $e(\cdot)$ is nondecreasing, and therefore the analysis is set up for $\hat{\theta} > \theta$.

Corollary 4 *Incentive compatibility requires for verification schemes* $(e, a_l, f) \in F_I$ *that*

$$e(\hat{\theta}) = e(\theta) = 0 \textit{ for } \theta < \hat{\theta} \leq l, \tag{4.4}$$

$$e(\hat{\theta}) - e(\theta) \leq f(\hat{\theta},\theta) \textit{ for } l < \theta < \hat{\theta}, \tag{4.5}$$

$$\textit{and} \quad e(\hat{\theta}) \leq f(\hat{\theta},l) \textit{ for } \theta \leq l < \hat{\theta}. \tag{4.6}$$

Proof: (4.4) and (4.5) follow directly from the incentive compatibility condition $\pi(\theta,\theta) \geq \pi(\hat{\theta},\theta)$. In addition firms with a true emissions value $\theta \leq l$ may prefer to announce a value $\hat{\theta} > l$, unless

$$\pi(\theta,\theta) = 0 \geq e(\hat{\theta}) - f(\hat{\theta},\theta) = \pi(\hat{\theta},\theta) \tag{4.7}$$

$$\Leftrightarrow f(\hat{\theta},\theta) \geq e(\hat{\theta}) \quad \forall\, \theta \leq l < \hat{\theta}. \tag{4.8}$$

Since the penalty function is nondecreasing in the degree of misreporting, for any given $\hat{\theta} > l$ and for $\theta \leq l$ the penalty is smallest if $\theta = l$ and therefore (4.8) holds if (4.6) holds. ■

Turning to *efficiency* and applying similar arguments as for random auditing, it follows from Proposition 1 that:

Corollary 5 *A mechanism $(e, a_l, f) \in F_I$ is efficient in F_I only if for a given penalty function the allowances are allocated such that*

$$\sup_{\theta<\hat{\theta}} \frac{e(\hat{\theta}) - e(\theta)}{f(\hat{\theta}, \theta)} = \begin{cases} 1 & \text{if } \hat{\theta} > l \\ 0 & \text{if } \hat{\theta} \leq l. \end{cases} \tag{4.9}$$

Thus, since efficient mechanisms are in particular incentive compatible, in efficient mechanisms the allocation function fulfills $e(\hat{\theta}) = 0$ for $\hat{\theta} \leq l$. For $\hat{\theta} > l$ the value of the allocation function is less straightforward determined. If the supremum of $\frac{e(\hat{\theta})-e(\theta)}{f(\hat{\theta},\theta)}$ over all $\theta < \hat{\theta}$ were attained at some $\theta \leq l$, efficiency would require $e(\hat{\theta}) = f(\hat{\theta}, l)$ from (4.6) and since

$$1 = \sup_{\theta\leq l<\hat{\theta}} \frac{e(\hat{\theta}) - e(\theta)}{f(\hat{\theta}, \theta)} = \sup_{\theta\leq l<\hat{\theta}} \frac{e(\hat{\theta})}{f(\hat{\theta}, \theta)} \overset{(2.8)}{=} \frac{e(\hat{\theta})}{f(\hat{\theta}, l)}, \tag{4.10}$$

using $e(\theta) = 0$ for all $\theta \leq l$. However, it can not be guaranteed, that the supremum is not attained at some $\theta > l$. Therefore the analysis is performed for the specific penalty structures introduced above.

Since the audit probability is one for all reports above the cutoff level, deterministic auditing resembles strongly the "Full Verification" scenario under random auditing, with an audit probability of one for every report. The question is, how the difference in the audit probabilities for reports below the cutoff level carries over to the allocation function of an efficient mechanism.

4.1.1 Proportional Penalty

For the proportional penalty $f_{prop}(\hat{\theta}, \theta) = \max\{0, \gamma(\hat{\theta} - \theta)\}$ the efficiency condition (4.9) is given by

$$\sup_{\theta<\hat{\theta}} \frac{e(\hat{\theta}) - e(\theta)}{\gamma(\hat{\theta} - \theta)} = \begin{cases} 1 & \text{if } \hat{\theta} > l \\ 0 & \text{if } \hat{\theta} \leq l. \end{cases} \tag{4.11}$$

In the following analysis the result is established that:

Proposition 11 *A mechanism* $(e_l, a_l, f_{prop}) \in F_I$ *is efficient in* F_I*, only if the allocation function fulfills*

$$e_l(\hat{\theta}) = f_{prop}(\hat{\theta}, l) = \begin{cases} \gamma(\hat{\theta} - l) & \text{if } \hat{\theta} > l \\ 0 & \text{if } \hat{\theta} \leq l \end{cases} \tag{4.12}$$

The allocation function $e_l(\cdot)$ is presented in Figure 4.1 for different values of γ.

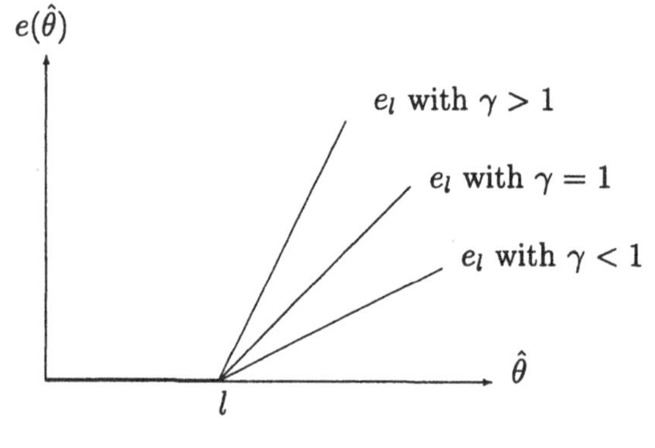

Figure 4.1: Efficient allocation function under proportional penalties

To see that (4.12) holds, it is used that $e(\hat{\theta}) = 0$ for all $\hat{\theta} \leq l$. It then has to be shown that $\gamma(\hat{\theta} - l)$ holds for all $\hat{\theta} > l$. For this, case (i): $\theta \leq l < \hat{\theta}$ is compared to case (ii): $l < \theta < \hat{\theta}$. In addition the analysis is performed separately for potentially concave and potentially convex allocation functions (with respect to the "upper part", i.e. $\hat{\theta} > l$). It will turn out that the only structure that holds in an efficient mechanism is the linear one which is given in (4.12).

In order to perform the analysis for allocation functions that are concave in the upper part, consider for illustrative purpose first a strictly concave allocation function with $e(0) = 0$, but not $e(\hat{\theta}) = 0$ for $\hat{\theta} \leq l$ and compare the following values:

$$(a) \ : \ \sup_{\theta \leq l < \hat{\theta}} \frac{e(\hat{\theta})}{\gamma(\hat{\theta}-\theta)} = \frac{e(\hat{\theta})}{\gamma(\hat{\theta}-l)} = \frac{e(\hat{\theta})}{f_{prop}(\hat{\theta},l)} \quad \forall\, \hat{\theta}, \tag{4.13}$$

since the supremum is attained at $\theta = l$.

$$(b) \ : \ \sup_{l<\theta<\hat{\theta}} \frac{e(\hat{\theta})-e(\theta)}{\gamma(\hat{\theta}-\theta)} = \frac{e(\hat{\theta})-e(l)}{\gamma(\hat{\theta}-l)} \quad \forall\, \hat{\theta}, \tag{4.14}$$

since the supremum is attained at $\theta \to l$.

For (b) it is used that $e(\cdot)$ is concave and continuous.

Figure 4.2 illustrates these cases for $\gamma = 1$ and also sets the suprema in relation to the efficient audit probability under random auditing (cf. Figure 3.1). The values of (4.13), (4.14) and the audit probability under random auditing can be read from the slope of the straight line through $(\hat{\theta}, e(\hat{\theta}))$ and respectively $(l, 0)$, $(l, e(l))$ and $(0, 0)$.

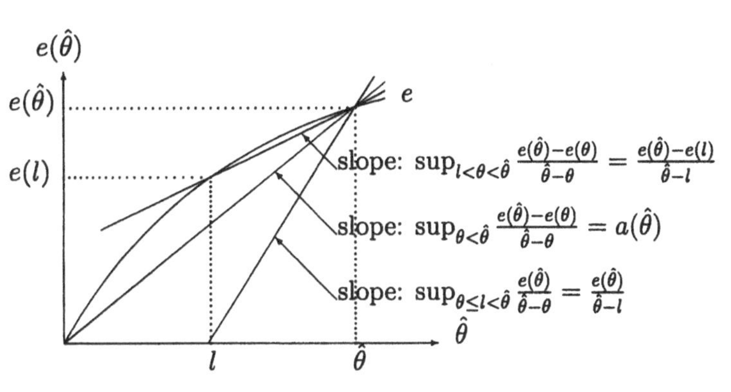

Figure 4.2: Illustration of (4.15) for $\gamma = 1$

It can be seen that in general

$$\begin{aligned} \sup_{\theta \leq l < \hat{\theta}} \frac{e(\hat{\theta})}{\gamma(\hat{\theta}-\theta)} = \frac{e(\hat{\theta})}{\gamma(\hat{\theta}-l)} &\geq \frac{e(\hat{\theta})}{\gamma\hat{\theta}} = \sup_{\theta<\hat{\theta}} \frac{e(\hat{\theta})-e(\theta)}{\gamma(\hat{\theta}-\theta)} = a(\hat{\theta}) \\ &\geq \sup_{l<\theta<\hat{\theta}} \frac{e(\hat{\theta})-e(\theta)}{\gamma(\hat{\theta}-\theta)} = \frac{e(\hat{\theta})-e(l)}{\gamma(\hat{\theta}-l)}. \end{aligned} \tag{4.15}$$

Going back to allocation functions with $e(\hat{\theta}) = 0$ for $\hat{\theta} \leq l$ and *concavity in the upper part*, it is seen that $\sup_{\theta \leq l < \hat{\theta}} \frac{e(\hat{\theta})}{\gamma(\hat{\theta}-\theta)}$ equals case (i) above and $\sup_{l<\theta<\hat{\theta}} \frac{e(\hat{\theta})-e(\theta)}{\gamma(\hat{\theta}-\theta)}$ case (ii). In addition, comparing Figure 4.2 and Figure 4.1 it comes clear that the three suprema coincide. Thus, (4.13) and (4.14) coincide and from (4.11) efficiency implies

$$\frac{e(\hat{\theta})}{\gamma(\hat{\theta}-l)} = 1 \quad \forall\, \hat{\theta} > l$$

$$\Longleftrightarrow \quad e(\hat{\theta}) = \gamma(\hat{\theta}-l). \tag{4.16}$$

An allocation function which is concave and not linear in the upper part can not be used to set up an efficient mechanism under deterministic auditing and proportional penalties.

Figure 4.3 also illustrates the relationship of case (i) and case (ii), this time for an allocation function that is *convex in the upper part.* It holds:

$$(ii) \;:\; \sup_{l<\theta<\hat{\theta}} \frac{e(\hat{\theta})-e(\theta)}{\gamma(\hat{\theta}-\theta)} = \frac{1}{\gamma} D^{-}e(\hat{\theta}) \quad \forall\, \hat{\theta}, \tag{4.17}$$

since the supremum is attained for $\theta \to \hat{\theta}$.

For e_l continuous and convex in the upper part, the comparison of (4.13) and (4.17) yields

$$\sup_{l<\theta<\hat{\theta}} \frac{e(\hat{\theta})-e(\theta)}{\gamma(\hat{\theta}-\theta)} = \frac{1}{\gamma} D^{-}e(\hat{\theta}) \geq \frac{1}{\gamma}\frac{e(\hat{\theta})}{\hat{\theta}-l} = \sup_{\theta \leq l < \hat{\theta}} \frac{e(\hat{\theta})}{\gamma(\hat{\theta}-\theta)} \quad \forall\, \hat{\theta} > l, \tag{4.18}$$

the slope of the secant through two points is less steep than the slope of the tangent at the upper point.

Applying again the efficiency condition (4.11), this implies for the upper part of an allocation function with $e(\hat{\theta}) = 0$ for $\hat{\theta} \leq l$:

$$\frac{1}{\gamma} D^{-}e(\hat{\theta}) = 1 \quad \forall\, \hat{\theta} > l$$

$$\Longleftrightarrow \quad e(\hat{\theta}) = \gamma\hat{\theta} + const. \quad \forall\, \hat{\theta} > l. \tag{4.19}$$

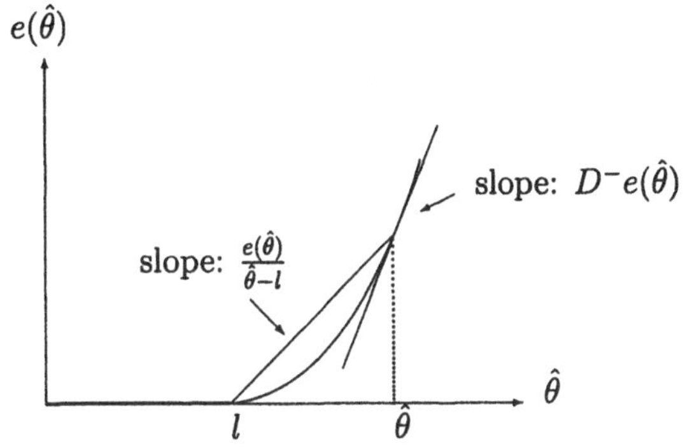

Figure 4.3: Illustration of (4.18)

Continuity of the function in l requires the constant to fulfill $\gamma l + const = 0 \Leftrightarrow const = -\gamma l$. i.e.

$$e_l(\hat{\theta}) = \gamma(\hat{\theta} - l). \tag{4.20}$$

Lust like the concave case, an allocation function which is convex and not linear in the upper part can not be used to set up an efficient mechanism under deterministic auditing and proportional penalties. Suppose continuity were not required and $const > -\gamma l$. In this case the condition $D^{-}e(\hat{\theta}) \geq e(\hat{\theta})/(\hat{\theta} - l)$ no longer holds, since:

$$D^{-}e(\hat{\theta}) \geq \frac{e(\hat{\theta})}{(\hat{\theta} - l)} \;\Leftrightarrow\; \gamma \geq \frac{\gamma\hat{\theta} + const}{\hat{\theta} - l} \tag{4.21}$$

$$\Leftrightarrow\; \gamma(\hat{\theta} - l) \geq \gamma\hat{\theta} + const \tag{4.22}$$

$$\Leftrightarrow\; -\gamma l \geq const. \tag{4.23}$$

This would imply for the upper part of the function:

$$\frac{e(\hat{\theta})}{\gamma(\hat{\theta} - l)} = 1 \Leftrightarrow e(\hat{\theta}) = \gamma(\hat{\theta} - l). \tag{4.24}$$

So the allocation function as given in Proposition 11 is obtained, whether continuity is required or not. Note that $e(\hat{\theta}) = \gamma(\hat{\theta} - l)$ has to hold *for all* $\hat{\theta} > l$, which rules out piecewise strictly concavity or convexity.

The very direct correlation between the penalty and the allocation function is immediately seen. Not surprisingly a higher penalty implies a higher allocation, and, the other way round, if the allocation shall be increased, the penalty has to be increased as well.

From Proposition 9 it is known that under random auditing and Full Verification the allocation function of an efficient mechanism is given by $e(\hat{\theta}) = \gamma\hat{\theta}$. The common ground of the allocation under Full Verification and under deterministic auditing is thus, that the allocation function of an efficient verification scheme is linear and uniquely determined by the audit probability and penalty parameter.

Remark 10 *For the same penalty, the allocation is lower under deterministic auditing than under random auditing. This is due to the fact that firms which report a value below the cutoff level will neither be audited nor receive any emission allowances. The difference between the allocation function under random auditing and under deterministic auditing is just the amount that refrains a type $\theta \leq l$ from misreporting.*

This amount is given by

$$e(\hat{\theta}) - e_l(\hat{\theta}) = \gamma\hat{\theta} - \begin{cases} \gamma(\hat{\theta} - l) & \text{if } \hat{\theta} > l \\ 0 & \text{if } \hat{\theta} \leq l \end{cases} = \begin{cases} \gamma l & \text{if } \hat{\theta} > l \\ \gamma\hat{\theta} & \text{if } \hat{\theta} \leq l, \end{cases} \tag{4.25}$$

and is equal to the amount a firm of type $\theta \leq l$ would maximally gain if under deterministic auditing the allocation function equalled $e(\hat{\theta}) = \gamma\hat{\theta}$:

$$\pi(\hat{\theta}, \theta) = \gamma\hat{\theta} - \begin{cases} \gamma(\hat{\theta} - \theta) & \text{if } \hat{\theta} > l \\ 0 & \text{if } \hat{\theta} \leq l \end{cases} = \begin{cases} \gamma\theta \leq \gamma l & \text{if } \theta \leq l < \hat{\theta} \\ \gamma\hat{\theta} & \text{if } \theta < \hat{\theta} \leq l. \end{cases} \tag{4.26}$$

This comparison has not yet taken into account that there is an upper limit on the overall amount of allowances. To see the effects of this additional constraint, consider the following scenario: Starting from an initial situation with Full Verification ($a(\cdot) \equiv 1$), proportional penalties ($f(\cdot) = f_{prop}(\cdot)$) and the associated efficient allocation function $e(\hat{\theta}) = \gamma\hat{\theta}$, where the penalty parameter γ is chosen such that all allowances are allocated. Suppose now, in order to save costs, the audit probability were changed to $a(\hat{\theta}) = 0$ for reports below a certain cutoff level l, and left to be one for reports above the cutoff level. The above analysis has shown, that efficiency requires, that now less allowances are allocated. Firms which report a value below the cutoff level will not receive any allowances, and firms which report a value above the cutoff level will receive $e(\hat{\theta}) = \gamma\hat{\theta} - \gamma l$, less than before. This has the consequence, however, that in aggregate less allowances are allocated, some are "spare". The allocation of all allowances can then be achieved, if γ is increased - the common parameter of the allocation and the penalty function. Since $e_l(\hat{\theta}) = 0$ for $\hat{\theta} \leq l$, the increase in γ ($=: \gamma'$) has to be such that $\gamma'(\hat{\theta} - l) > \gamma\hat{\theta}$ for all $\hat{\theta} > l$, i.e. the amount of allowances that a firm of type $\theta > l$ receives under deterministic auditing exceeds the amount that the firm would receive under Full Verification. Thus:

Remark 11 *A change from Full Verification to deterministic auditing has two major effects: firstly a redistribution of allowances from firms with low emission values to firms with high emission values, and secondly an increase in the size of the penalty.*

This holds for every cutoff level l, but obviously the effects are weaker for a threshold closer to zero.

4.1.2 Constant Penalty

For the constant penalty $f_{const}(\hat{\theta}, \theta) = \begin{cases} C & \text{if } \hat{\theta} > \theta \\ 0 & \text{if } \hat{\theta} \leq \theta \end{cases}$ the efficiency condition (4.9) is given by

$$\sup_{\theta < \hat{\theta}} \frac{e(\hat{\theta}) - e(\theta)}{C} = \frac{e(\hat{\theta})}{C} = a_l(\hat{\theta}) = \begin{cases} 1 & \text{if } \hat{\theta} > l \\ 0 & \text{if } \hat{\theta} \leq l. \end{cases} \tag{4.27}$$

Solving for $e(\cdot)$ yields

$$e_l(\hat{\theta}) = \begin{cases} C & \text{if } \hat{\theta} > l \\ 0 & \text{if } \hat{\theta} \leq l. \end{cases} \tag{4.28}$$

The value C is the penalty for all types θ and reports $\hat{\theta} > \theta$, in particular it is the penalty for type $\theta = l$. Interpreted like this, the allocation function can be written as in the proportional penalty case:

Corollary 6 *A mechanism $(e_l, a_l, f_{const}) \in F_I$ is efficient in F_I iff the allocation function fulfills*

$$e_l(\hat{\theta}) = f_{const}(\hat{\theta}, l). \tag{4.29}$$

The allocation function $e_l(\cdot)$ is displayed in Figure 4.4.

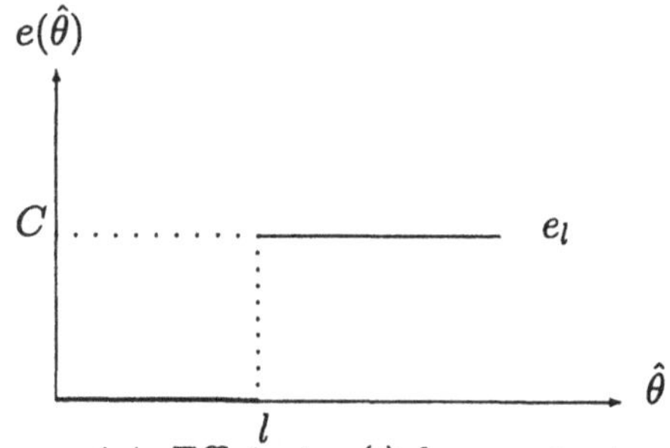

Figure 4.4: Efficient $e_l(\cdot)$ for constant penalties

Continuity can only be achieved for $e(\cdot) \equiv 0$, which of course is not feasible. Since under deterministic auditing the audit probability is fixed and the allocation and penalty function are determined by the constraint that a given amount of allowances has to be allocated, it is neither possible to decrease the audit probability, nor to increase the allocations. Thus, efficiency follows for noncontinuous allocation functions.

As in the case of Full Verification it can be shown that expected allowances are zero for all overstatements:

$$\pi(\hat{\theta}, \theta) = e_l(\hat{\theta}) - a_l(\hat{\theta}) f_{const}(\hat{\theta}, \theta) = \begin{cases} C - C = 0 & \text{if } l < \theta < \hat{\theta} \\ C - C = 0 & \text{if } \theta \leq l < \hat{\theta} \\ 0 - 0 = 0 & \text{if } \hat{\theta} \leq l. \end{cases} \tag{4.30}$$

The "driving forces" behind the results are the ones already explained in the chapter on Full Verification: the penalty is the same for all firms which have overstated their emissions. In addition the audit probability is fixed, therefore there is no scope for the determination of the allocation function and every firm with an emissions value above l has to receive the same amount of allowances.

4.1.3 Correct and Penalize

For a penalty that *corrects and penalizes*, $\tilde{f}(\hat{\theta},\theta) = \max\{0, (e(\hat{\theta}) - e(\theta) + f(\hat{\theta},\theta))\}$, the efficiency condition (4.9) is given by

$$\sup_{\theta<\hat{\theta}} \frac{e(\hat{\theta}) - e(\theta)}{e(\hat{\theta}) - e(\theta) + f(\hat{\theta},\theta)} = a_l(\hat{\theta}) = \begin{cases} 1 & \text{if } \hat{\theta} > l \\ 0 & \text{if } \hat{\theta} \leq l. \end{cases} \tag{4.31}$$

Note first, that deterministic auditing is never efficient when the function $f(\cdot)$ already induces incentive compatibility. This follows, just as in the case of Full Verification, from Proposition 8, which states that for the described scenario the efficient audit probability will be less than 1/2.

If the function $f(\cdot)$ does not induce incentive compatibility, the allocation function will be zero for reports below the cutoff level, as above. For reports above the cutoff level, (4.31) implies

$$\frac{e(\hat{\theta}) - e(\theta)}{e(\hat{\theta}) - e(\theta) + f(\hat{\theta},\theta)} \leq 1 \quad \forall\, \theta < \hat{\theta}. \tag{4.32}$$

This, however, is always true, independent of the function $e(\cdot)$, since $f(\hat{\theta},\theta) \geq 0$ for all $\hat{\theta} > \theta$ from feasibility condition (2.8). It holds for a true emissions value $\theta > l$, as well as $\theta \leq l$, since for the latter $e(\theta) = 0$ is plugged into the nominator as well as the denominator. It follows:

Corollary 7 *Let $(e_l, a_l, \tilde{f}) \in \tilde{F}_I$ be an efficient mechanism in $\tilde{F}_I$. The allocation function is given by*

$$e_l(\hat{\theta}) = \begin{cases} \textit{arbitrary} & \textit{if } \hat{\theta} > l \\ 0 & \textit{if } \hat{\theta} \leq l. \end{cases} \tag{4.33}$$

If the correcting penalty consists only of the gain from misreporting,

$$\tilde{f}(\hat{\theta},\theta) = \max\{0, (e(\hat{\theta}) - e(\theta))\}, \tag{4.34}$$

the mechanism $(e_l, a_l, \tilde{f}) \in \tilde{F}_I$ is efficient in $\tilde{F}_I$.

The results follow again from the results under Full Verification. Note that efficiency is determined in the upper part of the allocation function, since only allocation functions with $e(\hat{\theta}) = 0$ for $\hat{\theta} \leq l$ are feasible. Thus, just as under Full Verification, there is no need for an additional penalty part, it is sufficient to ensure that there is no gain. This relation also holds true the other way round: whenever the penalty does not exceed the gain from misreporting, an audit probability of one is necessary to induce incentive compatibility.

4.2 Efficiency in F

Noteworthy, efficiency in F excludes incentive compatibility for a whole class of penalty functions, as is seen below. For mechanisms element F, condition (4.3) holds. For any given l the audit probability is fixed and therefore efficiency is determined in the upper part of the allocation function. For any mechanism $(e, a_l, f) \in F$ with $e(\hat{\theta}) < f(\hat{\theta}, 0)$ for some $\hat{\theta} > l$, a mechanism $(e', a_l', f') \in F$ can be found with $e'(\cdot) \geq e(\cdot)$, $e'(\cdot) \neq e(\cdot)$ $a_l'(\cdot) = a_l(\cdot)$, and $f'(\cdot) = f(\cdot)$. It follows:

Proposition 12 *A mechanisms $(e, a_l, f) \in F$ is efficient in F iff*

$$e_{lF}(\hat{\theta}) = \begin{cases} f(\hat{\theta}, 0) & \text{if } \hat{\theta} > l \\ 0 & \text{if } \hat{\theta} \leq l \end{cases} \tag{4.35}$$

holds.

The proof is given in Appendix D.

Incentive compatibility may, however, not hold. Firms with an emissions value below the cutoff level will not necessarily report their true emissions value. The intuition for this result is as follows: firms with $\theta \leq l$ expect to receive no allowances when they report their true values. However, when they report a value above l they will receive $\pi(\hat{\theta}, \theta) = e(\hat{\theta}) - f(\hat{\theta}, \theta) = f(\hat{\theta}, 0) - f(\hat{\theta}, \theta) \geq 0$. Although they know with certainty that their misreporting will be detected and punished, they gain by reporting wrong data if $f(\hat{\theta}, 0) > f(\hat{\theta}, \theta)$. This is naturally the case for all increasing penalty functions. It will result in a situation where every firm reports emissions above the cutoff level and will therefore have to be audited, leading to high audit costs. The result is summarized in

Proposition 13 *Efficient verification schemes* $(e_{lF}, a_l, f) \in F$ *with deterministic auditing and increasing penalty functions are not incentive compatible.*

The result is driven by the fact that $\pi(\theta, \theta) = 0$ for all $\theta \leq l$.

The outcomes of efficient mechanisms element F can be reproduced by an incentive compatible mechanism, if, applying the reasoning of the revelation principle, the determination of the audit probability of the respective mechanism (e'_{lF}, a'_l, f') is given by

$$a'_l(\hat{\theta}) := a_l(\alpha(\hat{\theta})) = \begin{cases} 1 & \text{if } \alpha(\hat{\theta}) > l \\ 0 & \text{if } \alpha(\hat{\theta}) \leq l, \end{cases} \tag{4.36}$$

where $\alpha(\hat{\theta})$ is an optimal report for a firm of type $\hat{\theta}$ under the mechanism (e_{lF}, a_l, f). So firms still trigger an audit with their report, but it's not the report itself that is compared with l, but the optimal report for such a type under the old mechanism. Setting $e'(\hat{\theta}) := \pi(\alpha(\hat{\theta}), \hat{\theta})$ it holds $\pi'(\theta, \theta) = e'(\theta) = \pi(\alpha(\theta), \theta)$, and incentive compatibility can be shown using the same calculations as applied for the proof of Lemma 1 in Appendix B. The question arises, however, whether the outcome is really desirable, since after all it may require that all firms are audited (recall from above that for some penalty functions it's an optimal strategy for each type to report values above the cutoff level).

As examples, consider the above introduced penalties:

For a *proportional penalty*

$$f_{prop}(\hat{\theta}, \theta) = \max\{0, \gamma(\hat{\theta} - \theta)\} \tag{4.37}$$

the allocation function that leads to efficiency in F is given by (cf. (4.35))

$$e_{lF}(\hat{\theta}) = \begin{cases} \gamma\hat{\theta} & \text{if } \hat{\theta} > l \\ 0 & \text{if } \hat{\theta} \leq l, \end{cases} \tag{4.38}$$

From Proposition 13 it is known, that the mechanism will not be incentive compatible, since the penalty function is increasing. The allocation function $e_{lF}(\cdot)$ is displayed in figure 4.5. Different to the allocation function in an efficient and incentive compatible mechanism (cf. Figure 4.1), this function is not continuous at l.

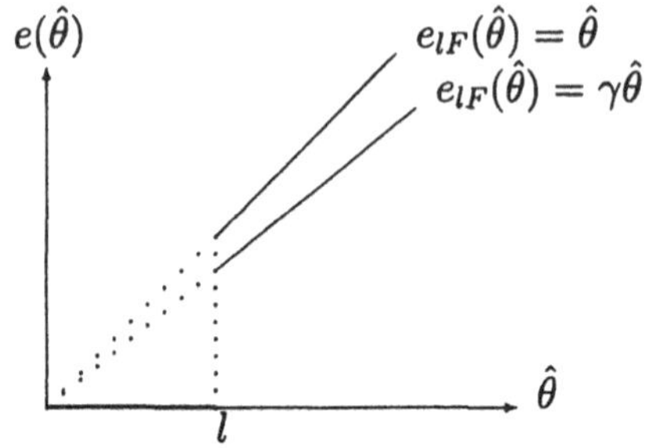

Figure 4.5: Efficiency in F for proportional penalties

Expected allowances are given by

$$\pi_{prop}(\hat{\theta},\theta) = \begin{cases} \gamma\hat{\theta} - \gamma(\hat{\theta}-\theta) = \gamma\theta & \text{if } \hat{\theta} > l \text{ and } \hat{\theta} > \theta \\ \gamma\hat{\theta} & \text{if } \hat{\theta} > l \text{ and } \hat{\theta} \leq \theta \\ 0 & \text{if } \hat{\theta} \leq l. \end{cases} \tag{4.39}$$

Since $\gamma\theta > \gamma\hat{\theta}$ for $\theta > \hat{\theta}$ firms will not underreport. Likewise $\gamma\theta > 0$, so, as stated above, every firm will report a value above l, every firm will have to be audited, and every firm will receive an amount of allowances equal to $\gamma\theta$. Although the set-up of the mechanism is different, it replicates the Full Verification scenario.

The strength of the penalty has, however, no influence on the reporting behavior of the firms, since a stronger penalty also implies higher allocations.

For a *constant penalty*

$$f_{const}(\hat{\theta},\theta) = \begin{cases} C & \text{if } \hat{\theta} > \theta \\ 0 & \text{if } \hat{\theta} \leq \theta \end{cases} \tag{4.40}$$

the allocation function that leads to efficiency in F is given by

$$e_{lF}(\hat{\theta}) = \begin{cases} C & \text{if } \hat{\theta} > l \\ 0 & \text{if } \hat{\theta} \leq l. \end{cases} \tag{4.41}$$

Incentive compatibility is achieved, since this allocation function is equal to the one given by (4.28), that leads to efficiency in F_I under a constant penalty.

Again, the reporting behavior of the firms is not dependent on the strength of the penalty, i.e. on the value of C.

For a penalty that *corrects and penalizes,*

$$\tilde{f}(\hat{\theta},\theta) = \max\{0, (e(\hat{\theta}) - e(\theta) + f(\hat{\theta},\theta))\} \tag{4.42}$$

the allocation function that leads to efficiency in $\tilde{F}$ is as well zero for reports below the cutoff level. For values above l the condition $\tilde{\pi}(\hat{\theta},0) \leq 0$ is always fulfilled: $e(\hat{\theta}) - \tilde{f}(\hat{\theta},0) = e(\hat{\theta}) - (e(\hat{\theta}) - e(0) + f(\hat{\theta},0)) = -f(\hat{\theta},0) \leq 0$ holds. Below it is shown that the sets $\tilde{F}$ and $\tilde{F}_I$ coincide. The allocation functions that lead to efficiency in $\tilde{F}$ are therefore those that lead to efficiency in $\tilde{F}_I$ (see (4.33)):

$$\tilde{e}_{lF}(\hat{\theta}) = \begin{cases} \textit{arbitrary} & \text{if } \hat{\theta} > l \\ 0 & \text{if } \hat{\theta} \leq l. \end{cases} \tag{4.43}$$

Note here the difference between the standard framework and the "tilde" framework, where Proposition 13 does no longer hold. To see that incentive compatibility holds although $\tilde{f}(\cdot)$ is increasing in the reported parameter, consider the expected allowances, which are given by

$$\tilde{\pi}(\hat{\theta},\theta) = \begin{cases} e(\theta) - f(\hat{\theta},\theta) < e(\theta) & \text{if } l < \theta < \hat{\theta} \text{ or } \theta \leq l < \hat{\theta} \\ e(\hat{\theta}) & \text{if } l < \hat{\theta} \leq \theta \\ 0 & \text{if } \hat{\theta} \leq l. \end{cases} \tag{4.44}$$

It can immediately be seen that such a penalty induces truth-telling by all firms, since expected penalties for overstating are always lower than those for truth-telling (keeping in mind that $e(\theta) = 0$ for $\theta \leq l$). If the cutoff level were the same, audit costs would be equal to those under deterministic auditing with constant penalties, but a penalty that corrects and penalizes possesses the advantage that the allocation function can be chosen freely for values above the cutoff level. In particular it can be differentiated with respect to the emissions of the firms.

Again, the parameter γ from the penalty function influences the allowances that are allocated to each firm. However, differently to proportional penalties, expected allowances are now decreasing in γ. The determining factor is the fact that the penalty is subtracted from $e(\theta)$, not $e(\hat{\theta})$.

4.3 Comparison of Audit Costs

In this section, deterministic and random audit schemes are compared with respect to total expected audit costs. It is assumed that the penalty function is the same in both schemes and that allowances are allocated proportionally.

For ease of computability let $f(\cdot) = f_{prop}(\hat{\theta}, \theta) = \max\{0, \gamma(\hat{\theta} - \theta)\}$. From above it follows for the allocation function under deterministic auditing:

$$e_l(\hat{\theta}) = f_{prop}(\hat{\theta}, l) = \begin{cases} \gamma(\hat{\theta} - l) & \text{if } \hat{\theta} > l \\ 0 & \text{if } \hat{\theta} \leq l. \end{cases}$$

The allocation function under random auditing is given by:

$$e(\hat{\theta}) = \lambda\hat{\theta} \quad \forall\, \hat{\theta} \in [0, \theta_u], \tag{4.45}$$

with $\lambda \in \mathbb{R}_+$. In the first part of the analysis it is assumed, that the cutoff level l and the parameter λ are such that the overall amount of allowances is kept in each of the schemes. Given these values, the conditions will be determined that favor one or the other auditing strategy in terms of audit costs. A second part will further analyze the values l and λ, and determine their relation to the cap on allowances, the penalty parameter γ and the influence on audit costs.

Total expected audit costs are given by

$$TC_r = \left(\int_0^{\theta_u} a(\theta) \cdot g(\theta)\, d\theta\right) \cdot n, \tag{4.46}$$

$$\begin{aligned} TC_d &= \left(\int_0^{l} 0 \cdot g(\theta)\, d\theta + \int_l^{\theta_u} 1 \cdot g(\theta)\, d\theta\right) \cdot n \qquad (4.47) \\ &= (1 - G(l)) \cdot n, \end{aligned}$$

for random and deterministic auditing respectively. The difference in audit costs is determined by the audit probability $a(\cdot)$, the probability distribution $G(\cdot)$, and the cutoff level l.

From Proposition 2 the audit probability for efficient random auditing is given by

$$a(\hat{\theta}) = \frac{\lambda}{\gamma} \quad \forall\, \hat{\theta} \in [0, \theta_u]. \tag{4.48}$$

This results in $TC_r = \frac{\lambda}{\gamma}n$. The *difference between audit costs* is then given by

$$\Delta := \left[\frac{\lambda}{\gamma} - (1 - G(l))\right] n. \tag{4.49}$$

Audit costs are equal for

$$\frac{\lambda}{\gamma} = 1 - G(l). \tag{4.50}$$

For $\frac{\lambda}{\gamma} > 1 - G(l)$ the audit costs under the random audit scheme will be higher than the costs under the deterministic audit scheme, and vice versa.

The concrete cost-equalizing cutoff-level can only be determined for specific distribution functions. Its relation to the audit probability under random auditing can, however, be determined more general:

Corollary 8 *The cost-equalizing cutoff level is decreasing in the audit probability:*

$$\frac{dl}{d(\lambda/\gamma)} < 0. \tag{4.51}$$

The result follows immediately from (4.50) and the monotonicity of the distribution function $G(\cdot)$. If the audit probability $\frac{\lambda}{\gamma}$ rises, audit costs for random auditing will increase. The same effect is achieved for deterministic auditing if the cutoff level is decreased.

A change in the audit probability (and also the cutoff level) may result from a change in the strength of the penalty, a change in the parameters of the distribution, or a change in the overall amount of allowances, since the latter two influence the "allocation parameter" λ and all three influence the cutoff-level l. For a further analysis it is therefore necessary to specify the values l and λ.

Consider first the influence of the total amount of allowances V on the random audit scheme. Let therefore $e(\hat{\theta}) = \lambda^v \hat{\theta}$ with λ^v such that

$$\int_0^{\theta_u} e(\theta)g(\theta)d\theta = \frac{V}{n} \tag{4.52}$$

$$\Leftrightarrow \quad \int_0^{\theta_u} \lambda^v \theta g(\theta)d\theta = \frac{V}{n} \tag{4.53}$$

$$\Leftrightarrow \quad \lambda^v = \frac{V}{n}[E_G(\theta)]^{-1}. \tag{4.54}$$

with

$$E_G(\theta) := \int_0^{\theta_u} \theta g(\theta) d\theta \tag{4.55}$$

the expected emissions value. For the allocation function $e(\hat{\theta}) = \lambda^v \hat{\theta}$ this implies

$$e(\hat{\theta}) = \frac{V}{n} \frac{\hat{\theta}}{E_G(\theta)}. \tag{4.56}$$

The first part determines an "equal share" of overall allowances for every firm. The second part allows for adjustments for differences in actual emissions. Assuming all firms reported truthfully, a firm with emissions equal to the expected value will receive exactly the equal share, firms with more (less) emissions will receive respectively more (less) allowances. In addition it holds:

Corollary 9 *The amount of allowances that each firm receives is increasing in the overall amount of allowances and decreasing in the number of firms, but it does not depend on the strength of the penalty:*

$$\frac{\partial \lambda^v}{\partial V} > 0, \quad \frac{\partial \lambda^v}{\partial n} < 0, \quad \text{and} \quad \frac{\partial \lambda^v}{\partial \gamma} = 0. \tag{4.57}$$

These results fulfill the expectations. A firm receives a "bigger piece of the cake", if either the cake is bigger, or if there are less people to share it with. A change in the penalty affects the audit probability, but not the allocation function.

The above mentioned changes in the exogenous variables affect total audit costs under random auditing in the following way:

Proposition 14 *Under the assumptions of this section, total expected random audit costs $TC_r = \frac{\lambda^v}{\gamma} n$ are decreasing in the strength of the penalty and constant in the number of firms:*

$$\frac{\partial TC_r}{\partial \gamma} < 0, \quad \frac{\partial TC_r}{\partial n} = 0. \tag{4.58}$$

In addition they react positively on changes in the total amount of allowances via the influence of λ^v:

$$\frac{\partial TC_r}{\partial \lambda^v} > 0, \quad \frac{\partial TC_r}{\partial V} > 0. \tag{4.59}$$

Results are easily seen from

$$TC_r = \frac{\lambda^v}{\gamma}n = \frac{V}{\gamma}[E_G(\theta)]^{-1}. \tag{4.60}$$

A higher penalty parameter leads to a lower audit probability for each firm, and thus to lower total expected audit costs. An increase of the number of firms has two opposed effects. More firms have to be audited, which increases total expected audit costs. On the other hand each firm receives less allowances, since the total number of allowances is fixed. This lowers the audit probability per firm. From (4.60) it can be seen, that these effects cancel out each other. An increase of the total amount of allowances V leads to an increase of the number of allowances per firm. If each firm receives more allowances, the audit probability per firm, and thus total expected audit costs, will increase, which explains the increase in audit costs for an increase in V.

Consider now the influence of the total amount of allowances V on the deterministic audit scheme. The cutoff level l^v that leads to the compliance with the total amount of allowances V fulfills

$$\int_0^{\theta_u} e_l(\theta)g(\theta)d\theta = \frac{V}{n} \tag{4.61}$$

$$\overset{(4.12)}{\Longleftrightarrow} \quad \int_0^l 0g(\theta)d\theta + \int_l^{\theta_u} \gamma(\theta - l)g(\theta)d\theta = \frac{V}{n}, \tag{4.62}$$

Applying the implicit function theorem (see Appendix D) yields

Corollary 10 *The cutoff level l^v is increasing in the size of the penalty, decreasing in the overall amount of allowances, and increasing in the number of firms:*

$$\frac{\partial l^v}{\partial \gamma} > 0, \ \frac{\partial l^v}{\partial V} < 0 \ \text{ and } \ \frac{\partial l^v}{\partial n} > 0. \tag{4.63}$$

Given (4.12) the results are intuitive. Since under deterministic auditing the allocation of allowances is directly connected to the penalty, a stronger penalty implies that every firm that receives allowances will receive more of them. However, as the overall amount of allowances is fixed, the number of firms that receive allowances has to be reduced, which is achieved by a rise in the cutoff level. On the other hand, an increase of the amount V of allowances under otherwise identical conditions (that imply in particular that allowances per firm stay the same) permits that more firms will receive allowances, which is the case if the cutoff level is decreased. An increase in the number of firms increases the number of firms that do not get audited, as well as the number of firms that get audited and receive allowances, which again conflicts with the fix overall amount of allowances and therefore requires an increase of the cutoff level.

The effect of changes in the exogenous variables on total expected audit costs can now be determined for deterministic auditing just as for random auditing:

Proposition 15 *Under the assumptions of this section, total expected deterministic audit costs TC_d are decreasing in the cutoff level and in the strength of the penalty, and increasing in the total amount of allowances:*

$$\frac{\partial TC_d}{\partial l^v} < 0, \quad \frac{\partial TC_d}{\partial \gamma} < 0, \quad \frac{\partial TC_d}{\partial V} > 0. \tag{4.64}$$

Calculations for all results are given in Appendix D. Less firms are audited when the cutoff level is higher, which reduces audit costs. As was shown above, an increase of the penalty increases the cutoff level, while an increase of total allowances decreases it, which leads to the respective changes of the audit costs. The effect of a change in the number of firms n on total audit costs is less straightforward. From above it is known that the cutoff level is increasing in the number of firms. Note however, that this may or may not increase total audit costs, depending on how far the cutoff level is increased and whether this compensates for the additional firms with an emissions type above the cutoff level. This indeterminableness can also be seen in

$$\frac{dTC_d}{dn} = \underbrace{\frac{\partial TC_d}{\partial l^v}\frac{\partial l^v}{\partial n}}_{<0} + \underbrace{\frac{\partial TC_d}{\partial n}}_{=(1-G(l^v))>0} . \tag{4.65}$$

Although it cannot explicitly be calculated from (4.65), it seems intuitive that total audit costs are increasing in the number of firms. If allocations per firm were unchanged, the increase of the cutoff level would not result in a change of total audit costs, since the number of firms that receive allowances, and therefore the number of firms that is audited, stays the same. The increase of the cutoff level has, however, a second effect: it reduces the allowances per firm, since they are equal to $\gamma(\theta - l)$ for every $\theta > l$ (cf. (4.12)). Less allowances per firm imply that more firms will receive allowances, the argument again being the fix amount of overall allowances V. Since the firms that receive allowances are exactly the firms that are audited, more firms are audited and thus these considerations indicate that total audit costs should increase.

A comparison of total expected audit costs under random and deterministic auditing shows that it can not be determined, whether changes in the penalty, number of firms, or allowances affect either of the schemes more. For the comparison, the difference in audit costs, $\Delta = \frac{\lambda}{\gamma} - (1 - G(l))$, is used. For $\Delta > 0$ the cutoff level l^v is greater than the cost-equalizing one and thus the audit costs for deterministic auditing are lower than those for random auditing. The reverse holds for $\Delta < 0$. It can then be shown that $\frac{\partial \Delta}{\partial \gamma}$, $\frac{\partial \Delta}{\partial V}$ and $\frac{\partial \Delta}{\partial n}$ are indetermined (see Appendix D).

4.4 Comparison with Random Auditing

Table 4.1 summarizes the forms that the efficiency condition takes up under the different audit and penalty structures. For each of the penalty functions the column "random auditing" gives the audit probability of an efficient mechanism under random auditing for arbitrary allocation functions, and the column "deterministic auditing" presents the allocation functions in an efficient mechanism under deterministic auditing, with

$$f_{prop}(\hat{\theta}, \theta) = \max\{0, \gamma(\hat{\theta} - \theta)\}, \qquad f_{const}(\hat{\theta}, \theta) = \begin{cases} C & \text{if } \hat{\theta} > \theta \\ 0 & \text{if } \hat{\theta} \leq \theta, \end{cases},$$

and $\tilde{f}(\hat{\theta}, \theta) = \max\{0, (e(\hat{\theta}) - e(\theta) + f(\hat{\theta}, \theta))\}$.

If deterministic auditing is the preferred policy, the penalty that corrects and penalizes seems to be the most favorable of the three considered, since it provides the biggest flexibility in the choice of the allocation function. Also under random auditing this penalty was shown to be advantageous, since the enlargement of the penalty (in the size of the undeserved gain from overstating) can better be motivated than a simple increase, and therefore this function can be considered to possess the biggest potential for saving

	random auditing	deterministic auditing
f_{prop}	$a(\hat{\theta}) = \frac{e(\hat{\theta})}{\gamma\hat{\theta}}$, e conc. $a(\hat{\theta}) = \frac{1}{\gamma}D^{-}e(\hat{\theta})$, e conv.	$e(\hat{\theta}) = f_{prop}(\hat{\theta}, l)$
f_{const}	$a(\hat{\theta}) = \frac{e(\hat{\theta})}{C}$	$e(\hat{\theta}) = f_{const}(\hat{\theta}, l)$
$\tilde{f}$	$a(\hat{\theta}) = \sup_{\theta<\hat{\theta}} \frac{e(\hat{\theta})-e(\theta)}{e(\hat{\theta})-e(\theta)+f(\hat{\theta},\theta)}$	$e(\hat{\theta}) = \begin{cases} \text{arbitrary} & \text{if } \hat{\theta} > l \\ 0 & \text{if } \hat{\theta} \leq l \end{cases}$

Table 4.1: Efficiency in F_I for random and deterministic auditing

audit costs. The audit probability is, however, slightly less easy to calculate and motivate.

While under deterministic auditing and proportional penalties a mechanism that is efficient in F is not incentive compatible, under the same penalty and random auditing all mechanisms that are efficient in F and have a concave (not necessarily strictly) allocation function are incentive compatible and are efficient in F_I as well.

The penalty function determines audit costs under random and under deterministic auditing via different channels. Under random auditing the penalty function determines the efficient audit probability and thus directly the audit costs. Under deterministic auditing the audit costs are determined by the cutoff level. For a given cutoff level the penalty function has therefore no influence on the audit costs. Taking the cap on allowances into account, however, reveals the link between the penalty function and the cutoff level. If not the cutoff level, but the penalty and allocation function are given, the cutoff level is determined by the overall amount of allowances, the distribution of emissions, and the allocation function. For proportional and constant penalties the allocation function is directly determined by the penalty function, which therefore indirectly determines the cutoff level. In case of the penalty that corrects and penalizes, where the penalty equals the expected gain from misreporting and the allocation function can be chosen freely in the upper part, the cutoff level is not directly dependent on the penalty any more and may be chosen for example based on equity or welfare aspects, or such that a given auditing budget is used. It has to be noted, however, that every firm of a type below the cutoff level will not receive any allowances. Thus a higher cutoff level does not only have the positive effect of saving audit costs, but it also has a stronger redistribution effect.

If the aim of the government were to obtain incentive compatibility given a fixed auditing budget, this could most easily be done by setting the cutoff level respectively. If in addition the penalty were chosen to simply equal the "non-deserved" gain from overstating, which seems easy to communicate, there is even freedom in the choice of the allocation function, which may be used to allow for political concessions. For a concrete determination of the cutoff-level that keeps a given auditing budget, it will be necessary to choose a specific distribution function. In the current context, the *truncated Pareto distribution* seems to suit best: a low percentage of the firms emits a high percentage of the emissions. This relationship can be seen in the diagrams in Appendix A. Such a relation was first identified for the distribution of income by PARETO (1897). In the meantime this distribution found its way into several fields of socio-economics and natural science, since many data sets are observed to follow it at least approximately. Numerous examples can be found with respect to such diverse fields as income distribution[1], groundwater hydrology[2], or earthquake magnitudes, forest fire areas, and oil and gas field sizes[3]. The truncated Pareto distribution, which is used in this analysis, has an upper, as well as a lower limit for the random variable.

The distribution function is given by[4]

$$G(\theta) = \begin{cases} 0 & (\theta \leq \theta_l) \\ 1 - \left(\left[1 - \tilde{\theta}^{\kappa}\right]^{-1} \left[(\frac{\theta_l}{\theta})^{\kappa} - \tilde{\theta}^{\kappa}\right]\right) & (\theta_l < \theta < \theta_u) \\ 1 & (\theta \geq \theta_u) \end{cases} \qquad (4.66)$$

with

$$\tilde{\theta} := \theta_l/\theta_u,$$

[1] See MANDELBROT (1960).
[2] See GROISMAN ET AL. (2004).
[3] Results are provided by BURROUGHS AND TEBBENS (2001), BURROUGHS AND TEBBENS (2002), and MALAMUD ET AL. (1998).
[4] See for example ABAN ET AL. (2006).

where θ_u is the upper and θ_l the lower limit ($\theta_l > 0$ for technical reasons). $\kappa > 0$, which is also known as *Pareto constant*, is the index of inequality of the distribution. The lower κ, the bigger the inequality.

Remark 12 *The cutoff level that leads to the fulfillment of a given auditing budget under the assumption of the truncated Pareto-distribution is given by*

$$
\begin{aligned}
TC_d =: B &= (1 - G(l))n && (4.67)\\
\Leftrightarrow G(l) &= 1 - \frac{B}{n} && (4.68)\\
\Leftrightarrow l^* &= \theta_l \left(\frac{B}{n} \left[1 - \tilde{\theta}^{\kappa} \right] + \tilde{\theta}^{\kappa} \right)^{-\frac{1}{\kappa}}. && (4.69)
\end{aligned}
$$

Calculations are given in Appendix D. Note that $\frac{B}{n} < 1$, since costs per audit are normalized to one and by assumption the audit budget of the government does not suffice to audit all firms.

Naturally, this cutoff-level is decreasing in the budget, and increasing in the number of firms:

$$\frac{\partial l^*}{\partial B} < 0 \quad and \quad \frac{\partial l^*}{\partial n} > 0. \tag{4.70}$$

If the government has a greater audit budget available, more firms can be audited, which is achieved by lowering the cutoff-level. This in turn reduces allocations per firm. An increase in the total number of firms for a fixed budget, on the other hand, requires to increase the cutoff level to hold the number of firms that are audited constant.

Remark 13 *The budget-fulfilling cutoff level reacts positively to changes in the parameters of the distribution function:*

$$\frac{\partial l^*}{\partial \theta_l} > 0, \quad \frac{\partial l^*}{\partial \theta_u} > 0, \quad \frac{\partial l^*}{\partial \kappa} > 0. \tag{4.71}$$

Calculations are given in Appendix D. An increase of θ_l or θ_u induces a shift of the support to the right, resembling the fact that there are more firms with high emissions. Therefore, for the same cutoff level, more firms will report emission values above this level and would have to be audited. So to keep the audit budget, the cutoff level has to increase. κ is the index of the inequality of the distribution. A higher κ therefore indicates a shift of the distribution to the right, which includes, just as the right-shift of the support does, that there will be more firms audited under deterministic auditing, unless the cutoff level is increased.

Whenever under deterministic auditing the cutoff level is set based on the audit budget and not the overall amount of allowances, the penalty and allocation function will have to be constructed such that this cap on allowances is met.

For the aim to obtain incentive compatibility given a fixed auditing budget, and the assumption that it is not desired that firms below the cutoff level receive no allowances and that therefore the random audit rule is chosen, more calculations are required to meet the auditing budget. In particular all interrelations between the penalty, allocation and audit function have to be taken into account.

In the extensive literature on income tax compliance, REINGANUM AND WILDE (1985) were the first authors who analyzed this issue in a framework with an audit probability that depends on the taxpayers reported income. They compare a random audit to an audit cutoff policy and find that *"random audit rules are weakly dominated by audit cutoff rules"*, and that *"given lump-sum taxes and fines, audit cutoff rules are the least-cost policies which induce truthful reporting of income"*. These findings result from two key assumptions. Firstly, random audits are, unlike in later publications and in this analysis, characterized by the fact that the probability of audit is a random number and independent of reported income. The first result may therefore be generalized to saying that audit rules that depend on the taxpayers' report are superior to those that don't. The other major difference to the model of this analysis is the lump-sum nature of taxes and fines in REINGANUM AND WILDE (1985), which turns out to be the driving factor behind their results that *"A necessary and sufficient condition for a policy*[5] $(p(\cdot), T, F)$ *to induce truthful reporting is that* $p(x) \equiv 1$ *for all* $x < T$*."* (Theorem 3, p.13) and that, under both auditing schemes, the penalty F is irrelevant. The above analysis has shown, that these results do not hold any more when allocations and fines are not lump-sum. As soon as taxes (in my analysis allowances) and fines are truly dependent on the firms' reports, the penalty has a strong

[5] $p(\cdot)$ is the audit probability function, T and F the lump-sum tax and fine respectively, and x the reported income.

influence on the results and the audit probability in efficient schemes is not necessarily a cutoff policy any more. Efficiency does, however, not exclude such schemes. This result was also established by CHANDER AND WILDE (1998), who state that *"efficient schemes may involve random or deterministic auditing"* and that in particular *"deterministic schemes are not precluded from being optimal."* An example of the latter within the analysis of the verification of firm data in a grandfathering process is given by choosing proportional penalties and the allocation function $e_l(\hat{\theta}) = \begin{cases} \gamma(\hat{\theta} - l) & \text{if } \hat{\theta} > l \\ 0 & \text{if } \hat{\theta} \leq l \end{cases}$, (cf. (4.12) or Figure 4.1), which will lead to an efficient verification scheme with deterministic auditing.

4.5 Summary

In this chapter, an audit cutoff policy was analyzed, in which firms trigger an audit when reported emissions exceed a given level. Results were compared to the random audit policy analyzed in the previous chapter, with respect to distributional effects as well as total expected audit costs.

The threshold in the audit structure of the deterministic audit scheme carries over to the allocation of allowances, implying that firms with an emissions value below the cutoff-level have to be supplied with zero allowances. Comparing such a scheme with Full Verification under random auditing shows that, due to the cap on allowances, deterministic auditing necessarily leads to a redistribution of allowances from low-emission firms to high-emission firms. The allocations for firms with an emissions value above the cutoff-level are determined by the penalty. If the penalty is increasing in the report, the allocations are as well, whereas firms above the cutoff level will all receive the same amount of allowances when the penalty is constant.

In general, an audit probability of one requires that there is no gain from an overstatement in order for a mechanism to be incentive compatible. This is either done by relating the allocation function directly to the penalty function, as in the case of the proportional or constant penalty, or by explicitly setting the penalty equal to the gain from misreporting, as in the case of the penalty that corrects and penalizes. This has the advantage, that no additional penalty has to be imposed, and in particular that the government is free to chose the allocation function.

The comparison of total audit costs for random and deterministic auditing depends on the penalty and allocation functions (which determine the audit probability for efficient random audits), the distribution of the emission values, and the cutoff-level. For the calculable scenario with a proportional allocation of allowances and proportional penalties, an increase of the penalty provides a possibility for the government to reduce total audit costs, since the individual audit probability is decreasing in the penalty and total audit costs are positively related to individual audit costs. From the two exogenous factors firm number and total amount of allowances, only the latter has an influence on total random audit costs. It would, however, be wrong to think that the auditing can generally be relaxed when more allowances are at hand. If the penalty stays the same, a higher allocation simply implies higher incentives to overstate and therefore has to go along with a higher audit probability, increasing total audit costs. In the number of firms, however, total audit costs are constant. This is due to the fact that the two opposing effects (an increase of the number of firms that have to be audited versus a decrease of the audit probability per firm) cancel out each other.

The results for total deterministic audit costs are similar: decreasing in the penalty and increasing in the total amount of allowances. The argument is only slightly different and depends on the interplay of in particular the penalty and allocation function on the one hand, and the cap on allowances and the cutoff-level, on the other hand. If the allocation per firm stays the same, more total allowances simply imply that more firms will receive some, which means that more firms have to be audited and total audit costs increase. A higher penalty, which leads to a lower individual audit probability under random auditing, has now the effect of a higher allocation function, which in turn, due to the cap on allowances, implies a rise of the cutoff-level and therefore a decrease of audit costs.

With respect to the penalty function, the penalty that corrects and penalizes may be considered to be the most advantageous. Under deterministic auditing, it provides policy makers an opportunity to freely choose the way allowances shall be allocated to firms with an emissions value above the cutoff-level, whereas the allocation function of an efficient mechanism is already determined under proportional or constant penalties. In a random audit scheme, the advantage of the penalty that corrects and penalizes lies in the possibility to save audit costs by increasing the penalty in a seemingly acceptable way.

If under deterministic auditing the requirement of incentive compatibility is dropped, firms with an emissions value below the cutoff-level actually gain by declaring a value above the cutoff-level, whenever the penalty function is chosen to be increasing, as for example the proportional penalty. Thus, such firms will overstate, although they know with certainty that their misreporting will be detected and punished. In particular this implies, that all firms will report a value above the cutoff-level and will have to be audited.

The set-up of a verification scheme with deterministic auditing is more straightforward than that of a scheme with random auditing, if the aim is to fulfill a given auditing budget. If, however, the structure for the allocation of the allowances is politically determined and provides a positive amount of allowances for every firm with a positive amount of emissions, a scheme with random auditing has to be implemented.

Chapter 5

Conclusion

The preceding analysis answers the question of how, in the grandfathering process of an emissions trading scheme, self-reported firm data can be verified in a cost-efficient way. The increasing awareness of the risks of climate change and the influence of greenhouse gases has lead to the implementation of a wide range of policy instruments over the last years, all aiming to reduce these gases on a global level. The assignment of tradeable emission rights is an instrument with well-established advantages. Still, the transformation from textbook to reality discloses difficulties, one of them being the self-reported character of all firm data and the subsequent necessity of costly verification. This leads to the core of the analyzed problem, the auditing of private information, and it becomes apparent that the results have to be imbedded into the field of environmental economics as well as informational asymmetries.

The challenges of the analysis result from this mixture of classification: Firstly, the aim is to answer a question arising from a current political problem, and to provide policy recommendations regarding the efficient verification of firm data in the grandfathering process of the initial stages of an emissions trading scheme. This requires the set-up of the model to fit concrete, realistic guidelines. One of those with a strong influence on the model setup is, for example, the requirement of no rewards for truth-telling, since it makes additional constraints necessary for the use of the revelation principle. Secondly, the model, as well as the results have to be imbedded in, and in particular distinguished from related fields of literature.

It was shown that, although the question to be answered arises in an emissions trading context, the analysis does not belong to the classical emissions trading literature; although a fixed amount of emission allowances is allocated to the firms, firms don't take their rivals' action into consideration and the problem is no rent-seeking problem; although the problem deals with data verification, it does not belong to the field of compliance and enforcement; and even although the problem parallels the problem of tax evasion, and a principal-agent model similar to one in this strand of literature is set up, the parallel is not complete and results prove to be different, due to the different direction of misreporting and to the different motivations leading to the definition of efficiency. The equity-efficiency tradeoff that is known in the tax evasion literature does not exist in the grandfathering context, where an efficient scheme may include any feasible allocation function. The two fields do not only differ in the number of efficient solutions, but also in the forms of each of them, since, as was shown, efficient verification of private data strongly depends on whether there are incentives to overstate or to understate the private data.

The analyzed problem arises when allowances are grandfathered free of charge, since then the amount of allowances that a firm initially receives is dependent on its own report of its historic emissions, and costly verification is therefore essential to prevent that firms have an advantage from misreporting their data. In order to achieve high verification standards at low costs, I recommend to combine the audits with penalties for overstatements, which will serve as a deterrence and will therefore allow for a reduction of audits. Not surprisingly, such an approach replicates the Becker solution. However, it can be assumed that a disproportionately high penalty is politically not feasible in the context of emissions trading. Therefore the impact of different penalty structures and of the need for high verification standards, given by incentive compatibility, on the efficient audit probability have been analyzed. In general, the audit probability should equal the largest possible ratio of gross gain to penalty for an overstatement. Any allocation function can be applied to set up an efficient verification scheme, if the audit probability and penalty function are chosen respectively, while the need for high verification standards does in some of the scenarios put quite strong restrictions on the set of feasible mechanisms.

The exact form and size of the penalty has to be decided elsewhere and is likely to be based on legal or fairness aspects. It was shown, however, that it is favorable to construct the penalty such that it can be interpreted to consist of two parts: one part that simply equals the gain that resulted if there were no audit, and a second, actual penalty part. A penalty that consists only of the first part, the correction for the gain of overstating, replicates the current system: the audit probability is one for every firm and no advantage occurs from an overstatement. Similarly, if the audit strategy is predetermined to be a 100% verification, efficiency can be achieved for any allocation function by setting the penalty equal to the gain from misreporting for every report. Increasing the penalty further by adding an additional penalty part will obviously reduce audits, and with this audit costs. Why not use only this second part then? As argued in the analysis above, a penalty that corrects for "undeserved" gains is more likely to be politically acceptable than a penalty that is simply very high.

The structure of the penalty part determines whether the audit probability is increasing or decreasing in the reported value. Constant penalties, for example, are the same for a small or a big overstatement. Thus, firms that overstate are more likely to do so to a great extend. This leads to a higher efficient audit probability for higher reports. Penalties that are proportional to and increasing in the amount overstated, on the other hand, punish large overstatements stronger than small ones. In that case it depends on the allocation of the allowances, whether the audit probability should be increasing or decreasing in the reported value. In general, a higher initial allocation for any emissions value requires stronger audits of the report of that value. A regressive allocation of allowances then requires only a lower audit probability for higher reports than for lower reports, while it is the other way round for a progressive allocation of allowances.

Due to the complex interrelation of the initial allowances, penalty and audit probability in efficient verification schemes, keeping a given audit budget is most easily achieved by an audit structure that audits firms with an emissions value above a certain threshold with probability one, and the firms below the threshold with probability zero. The "price" for this accuracy, however, is a redistribution of emission allowances from firms with low emission values to firms with high emission values, since incentive compatibility requires not to allocate any allowances to firms that are not audited. In particular this implies that audit costs are negatively related to redistribution and a reduction of the audit budget has the direct effect of a stronger redistribution, which may or may not be politically desired. In addition, in order to maintain incentive compatibility, penalties for the audited firms will have to be higher under such a scenario than under a scenario with Full Verification of all reports.

This suggests therefore, that such a scheme with deterministic auditing seems to be less favorable than a scheme with random auditing, despite of its straightforward audit structure and determination of audit costs.

Total expected audit costs depend on the penalty and allocation functions, as well as the distribution of emissions values and, in case of the deterministic audit scheme, the cutoff-level and they can, under random as well as deterministic auditing, be reduced by an increase of the penalty.

As always, conclusions depend on model assumption. Reasonable extensions of the model would surely be the inclusion of risk-averse agents or the distinction between government and verifier. A further interesting aspect is the consideration of additional information that might exist about the firms. It has to be noted, however, that this differs very much from making the model dynamic. The dynamic context would be applicable for example for the yearly inspections, where firms have to show that they hold a sufficient amount of allowances for their *current* emissions. In that case, firms are more likely to have incentives to *understate* their emissions and thus the tax evasion framework should be applied.

Regarding the structure of the analysis, it would be worthwhile to check in how far the model and results can be applied to other fields of literature and situations where individuals have private information and may have incentives to overstate their data.

Appendix A

Annex B

Party	Quantified emission limitation or reduction commitment (percentage of base year or period)
Australia	108
Austria	92
Belgium	92
Bulgaria*	92
Canada	94
Croatia*	95
Czech Republic*	92
Denmark	92
Estonia*	92
European Community	92
Finland	92
France	92
Germany	92
Greece	92
Hungary*	94
Iceland	110
Ireland	92
Italy	92
Japan	94
Latvia*	92
Liechtenstein	92
Lithuania*	92
Luxembourg	92
Monaco	92
Netherlands	92
New Zealand	100
Norway	101
Poland*	94
Portugal	92
Romania*	92
Russian Federation*	100
Slovakia*	92
Slovenia*	92
Spain	92
Sweden	92
Switzerland	92
Ukraine*	100
United Kingdom of Great Britain and Northern Ireland	92
United States of America	93

* Countries that are undergoing the process of transition to a market economy.

Figure A-1: Annex B to the Kyoto Protocol

Source: UNFCCC (1997)

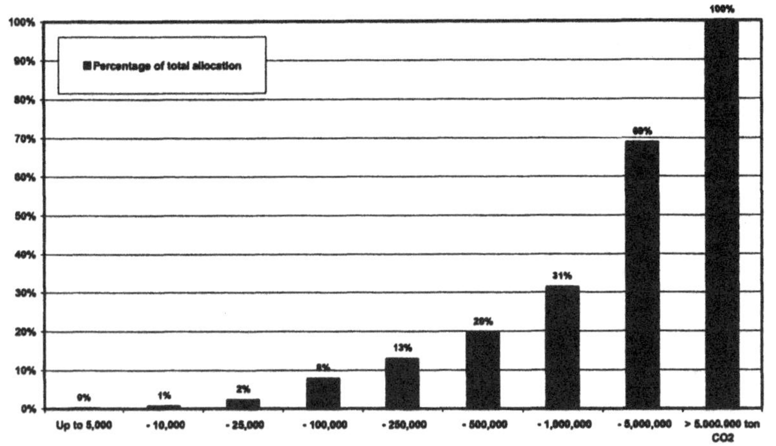

Figure 1 Cumulative amount of allocations per installation size class. Source of data: initially notified NAPsI as made available on the Commission website

Figure A-2: Distribution of emissions across firms in Europe (1)

Source: EC AND ECOFYS (2007)

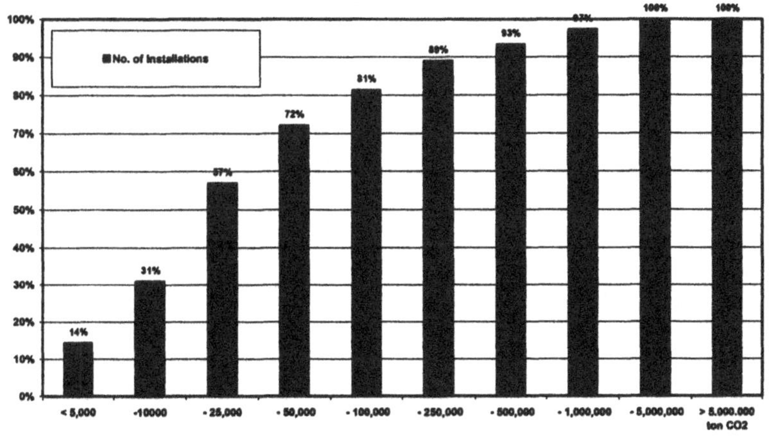

Figure 2 Cumulative amount of installations per installation size class. Source of data: initially notified NAPsI as made available on the Commission website

Figure A-3: Distribution of emissions across firms in Europe (2)

Source: EC AND ECOFYS (2007)

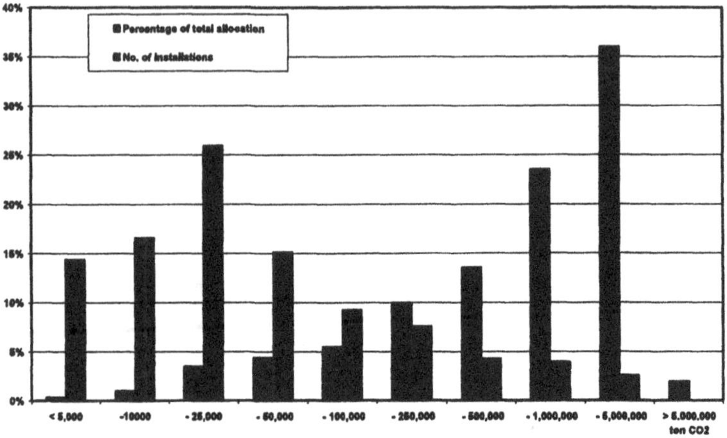

Figure 3 Number of installations and corresponding allocations per installation size class. Source of data: initially notified NAPsI as made available on the Commission website

Figure A-4: Distribution of emissions across firms in Europe (3)

Source: EC AND ECOFYS (2007)

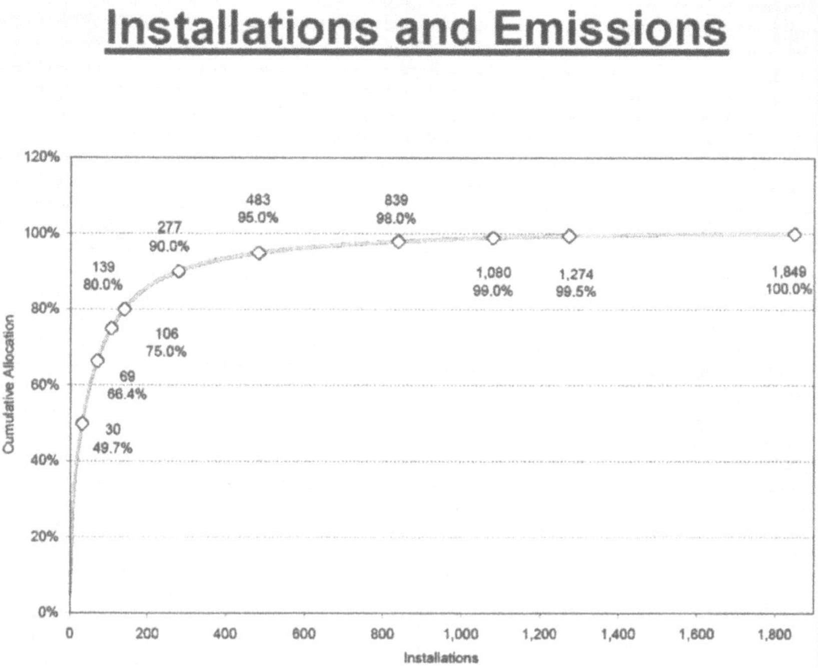

Figure A-5: Distribution of emissions across firms in Germany (1)

Source: UMWELTBUNDESAMT (2005)

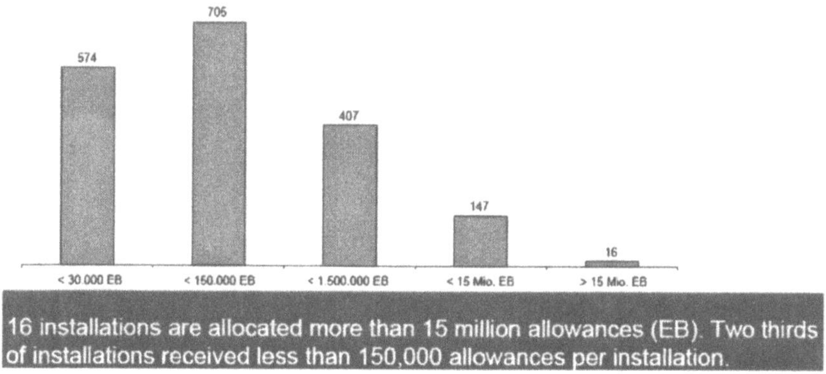

Figure A-6: Distribution of emissions across firms in Germany (2)

Source: DEHST (2005)

Appendix B

Proof of Lemma 1:
Let $(e, a, f) \in F$. Define (e', a', f') from (e, a, f) as follows:

$$\begin{aligned} e'(\hat{\theta}) &:= e\left(\alpha(\hat{\theta})\right) - a\left(\alpha(\hat{\theta})\right) f\left(\alpha(\hat{\theta}), \hat{\theta}\right) = \pi(\alpha(\hat{\theta}), \hat{\theta}), \\ a'(\hat{\theta}) &:= a\left(\alpha(\hat{\theta})\right), \\ f'(\hat{\theta}, \theta) &:= f(\hat{\theta}, \theta), \end{aligned}$$

where $\alpha(\theta), (\alpha(\theta) \geq \theta)$, is an optimal report for a firm of type θ under the mechanism (e, a, f). So the initial allocation and audit probability of the new mechanism (e', a', f') are given respectively by the expected allowances and audit probability of the old mechanism (e, a, f). The function values for the report $\hat{\theta}$ under the new mechanism thereby correspond to the values that would have resulted under the old mechanism if $\hat{\theta}$ had been the true value and if the firm had reported its optimal value $\alpha(\hat{\theta})$ given $\hat{\theta}$. Note that the penalty function is exactly the same in both mechanisms. In chapter 2.2 it was shown via the mechanism $(e'(\hat{\theta}), a'(\hat{\theta}), f'(\hat{\theta}, \theta)) := (e(\alpha(\hat{\theta})), a(\alpha(\hat{\theta})), f(\alpha(\hat{\theta}), \theta))$, that mechanisms which change the penalty function may lead to problems, since they don't fix the penalty function to zero at the point of the truthful report. Therefore in the mechanism that is used for the proof of the revelation principle in this analysis the penalty function is taken over and instead the allocation function is changed.

(a) Show that $(e', a', f') \in F$:

   (i) $e' : [0, \theta_u] \rightarrow \mathbb{R}_+$ holds, since it equals the expected allocation for a firm of type θ under the mechanism (e, a, f) if it pronounces $\alpha(\theta)$. This is not smaller than 0, since $e(\hat{\theta}) \geq 0$ for all $\hat{\theta}$ and the firm can secure itself a punishment of 0 if it pronounces the true type θ. As $\alpha(\theta)$ is an optimal report it holds $e'(\theta) \geq 0$.

(ii) $e'(0) = 0$ since $\pi(\alpha(0), 0) = 0$ due to condition (2.10), $\pi(\hat{\theta}, 0) \leq 0$, and the optimality of $\alpha(\cdot)$.

(iii) $e(\cdot) \not\equiv 0$ induces $\pi(\alpha(\hat{\theta}), \hat{\theta}) \not\equiv 0$ and therefore $e'(\cdot) \not\equiv 0$. The argument is again that $\alpha(\cdot)$ is an optimal report and therefore $\pi(\alpha(\hat{\theta}), \hat{\theta}) \geq e(\hat{\theta})$ has to hold.

(iv) $\pi'(\hat{\theta}, 0) = e'(\hat{\theta}) - a'(\hat{\theta})f'(\hat{\theta}, 0) = \pi(\alpha(\hat{\theta}), 0) - a(\alpha(\hat{\theta}))f(\hat{\theta}, 0) \leq 0$, since $\pi(\alpha(\hat{\theta}), 0) \leq 0$ and $a(\alpha(\hat{\theta}))f(\hat{\theta}, 0) \geq 0$.

In addition $0 \leq a'(\hat{\theta}) \leq 1$ and the conditions regarding $f(\cdot)$ carry over to $f'(\cdot)$ and thus $(e', a', f') \in F$.

(b) Show that $e'(\cdot)$ is nondecreasing:
Let $\theta^1 > \theta^2$, show that $e'(\theta^1) \geq e'(\theta^2)$:

$$\begin{aligned} e'(\theta^2) &= e(\alpha(\theta^2)) - a(\alpha(\theta^2))f(\alpha(\theta^2), \theta^2) \\ &\leq e(\alpha(\theta^2)) - a(\alpha(\theta^2))f(\alpha(\theta^2), \theta^1) \\ &\leq e(\alpha(\theta^1)) - a(\alpha(\theta^1))f(\alpha(\theta^1), \theta^1) \\ &= e'(\theta^1). \end{aligned}$$

To see why this holds note that due to $\theta^1 > \theta^2$ and $f(\cdot)$ nondecreasing in the degree of misreporting it holds that $f(\alpha(\theta^2), \theta^2) \geq f(\alpha(\theta^2), \theta^1)$ for both $\alpha(\theta^2) > \theta^1$ and $\alpha(\theta^2) \leq \theta^1$. In addition $\alpha(\theta^1)$ is optimal for θ^1 and therefore the second inequality holds.

(c) Show that (e', a', f') is incentive compatible:
The mechanism is incentive compatible if expected allowances are maximized for reporting the true emissions value, i.e. if

$$\pi'(\theta, \theta) \geq \pi'(\hat{\theta}, \theta) \quad \forall\, \hat{\theta}, \theta \in [0, \theta_u].$$

This is equivalent to

$$e'(\theta) \geq e'(\hat{\theta}) - a'(\hat{\theta})f'(\hat{\theta}, \theta) \quad \forall\, \hat{\theta}, \theta \in [0, \theta_u].$$

However, since $f'(\hat{\theta}, \theta) = f(\hat{\theta}, \theta) = 0$ for all $\hat{\theta} \leq \theta$ and since $e(\cdot)$ is nondecreasing as shown above, no firm will have an incentive to report a parameter $\hat{\theta} < \theta$. It is therefore now sufficient to show that

$$e'(\theta) \geq e'(\hat{\theta}) - a'(\hat{\theta})f'(\hat{\theta}, \theta) \quad \forall\, \hat{\theta} > \theta.$$

Let $\hat{\theta} > \theta$, then $e'(\theta) \geq e'(\hat{\theta}) - a'(\hat{\theta})f'(\hat{\theta}, \theta)$ is equivalent to

$$\begin{aligned} & e(\alpha(\theta)) - a(\alpha(\theta))f(\alpha(\theta), \theta) \\ & \quad \geq e(\alpha(\hat{\theta})) - a(\alpha(\hat{\theta}))f(\alpha(\hat{\theta}), \hat{\theta}) - a(\alpha(\hat{\theta}))f(\hat{\theta}, \theta) \\ & \quad = e(\alpha(\hat{\theta})) - a(\alpha(\hat{\theta}))[f(\alpha(\hat{\theta}), \hat{\theta}) + f(\hat{\theta}, \theta)]. \end{aligned}$$

In order to show that this holds, the fact that $\alpha(\theta)$ is an optimal report for a firm with emission parameter θ, and condition (2.11), $f(\alpha(\hat{\theta}), \theta) \leq f(\alpha(\hat{\theta}), \hat{\theta}) + f(\hat{\theta}, \theta)$, are used:

$$\begin{aligned} & e(\alpha(\theta)) - a(\alpha(\theta))f(\alpha(\theta), \theta) \\ & \quad \geq e(\alpha(\hat{\theta})) - a(\alpha(\hat{\theta}))f(\alpha(\hat{\theta}), \theta) \\ & \quad \geq e(\alpha(\hat{\theta})) - a(\alpha(\hat{\theta}))[f(\alpha(\hat{\theta}), \hat{\theta}) + f(\hat{\theta}, \theta)]. \end{aligned}$$

(d) Show that (e', a', f') replicates the equilibrium outcome arising from (e, a, f):

  (i) The outcome for the government is the audit it has to perform and therefore given by the audit probability. This is the same under the two mechanisms from the definition of $a'(\hat{\theta})$.

  (ii) The outcome for the firm is the amount of expected allowances. This is the same under the two mechanisms since

$$\begin{aligned} \pi(\alpha(\theta), \theta) &= e(\alpha(\theta)) - a(\alpha(\theta))f(\alpha(\theta), \theta) \\ &= e'(\theta) = e'(\theta) - a'(\theta)f'(\theta, \theta) \\ &= \pi'(\theta, \theta), \end{aligned}$$

due to $f'(\theta, \theta) = 0$. The amount of allowances that the firm expects to receive under the new mechanism (e', a', f') when it tells the truth equals just the amount of allowances that it expected to receive under the old mechanism (e, a, f) and the announcement of the optimal (i.e. expected allowances maximizing) report $\alpha(\theta)$.

■

Proof of Proposition 1:[1]

1. Let $(e, a, f) \in F_I$ be efficient in F_I. Since the mechanism is incentive compatible, (2.20) holds:

$$a(\hat{\theta}) \geq \sup_{\theta<\hat{\theta}} \frac{e(\hat{\theta}) - e(\theta)}{f(\hat{\theta}, \theta)} \quad \forall\, \hat{\theta}.$$

If the incentive constraint is not binding, i.e. if there exist some $\hat{\theta}$ with

$$a(\hat{\theta}) > \sup_{\theta<\hat{\theta}} \frac{e(\hat{\theta}) - e(\theta)}{f(\hat{\theta}, \theta)},$$

the inequality still holds, and therefore incentive compatibility is still provided, if the audit probability $a(\hat{\theta})$ is reduced, while $e(\cdot)$ and $f(\cdot)$ are not changed. However, this contradicts the assumption that (e, a, f) is efficient in F_I and therefore

$$a(\hat{\theta}) = \sup_{\theta<\hat{\theta}} \frac{e(\hat{\theta}) - e(\theta)}{f(\hat{\theta}, \theta)} \quad \forall\, \hat{\theta}$$

has to hold.

2. Let $(e, a, f) \in F_I$ with $a(\hat{\theta}) = \sup_{\theta<\hat{\theta}} \frac{e(\hat{\theta}) - e(\theta)}{f(\hat{\theta}, \theta)}$ for all $\hat{\theta}$. Show: $(e, a, f) \in F_I$ is efficient in F_I.
Suppose $(e, a, f) \in F_I$ is not efficient in F_I. Then there exists a mechanism $(e', a', f') \in F_I$ with $f' = f$ and

$$\begin{aligned} & a'(\cdot) \leq a(\cdot),\ a'(\cdot) \neq a(\cdot) \text{ and } e'(\cdot) \geq e(\cdot), \\ \textit{or} \quad & e'(\cdot) \geq e(\cdot),\ e'(\cdot) \neq e(\cdot) \text{ and } a'(\cdot) \leq a(\cdot). \end{aligned}$$

(a) Let $e'(\cdot) = e(\cdot)$. It follows that $a'(\cdot) \geq a(\cdot)$, since $(e', a', f') \in F_I$:

$$a'(\hat{\theta}) \geq \sup_{\theta<\hat{\theta}} \frac{e'(\hat{\theta}) - e'(\theta)}{f'(\hat{\theta}, \theta)} = \sup_{\theta<\hat{\theta}} \frac{e(\hat{\theta}) - e(\theta)}{f(\hat{\theta}, \theta)} = a(\hat{\theta}).$$

Thus, there is no $(e', a', f') \in F_I$ with $e'(\cdot) = e(\cdot)$, $a'(\cdot) \leq a(\cdot)$, $a'(\cdot) \neq a(\cdot)$.

[1] I am indebted to W. Richter for valuable comments in the trickier part of this proof.

(b) Let $e'(\cdot) \geq e(\cdot)$, $e'(\cdot) \neq e(\cdot)$, i.e. $e'(\hat{\theta}) > e(\hat{\theta})$ for some $\hat{\theta}$. The aim is to show that for this $\hat{\theta}$ it holds

$$a'(\hat{\theta}) \geq \sup_{\theta<\hat{\theta}} \frac{e'(\hat{\theta}) - e'(\theta)}{f'(\hat{\theta}, \theta)} > \sup_{\theta<\hat{\theta}} \frac{e(\hat{\theta}) - e(\theta)}{f(\hat{\theta}, \theta)} = a(\hat{\theta}).$$

This is done by showing that

$$e'(\hat{\theta}) - e'(\theta) > e(\hat{\theta}) - e(\theta) \quad \forall\, \theta < \hat{\theta}.$$

Let $\tilde{x}$ be the smallest value for which $e'(x) = e(x)$ for all $x \in [0, \tilde{x}]$ and $e'(\tilde{x}+\epsilon) > e(\tilde{x}+\epsilon)$ for $\epsilon > 0$. Such a value exists, since $e'(0) = e(0) = 0$. Let $\tilde{\epsilon} > 0$ be the value for which $e'(\tilde{x} + \epsilon) > e(\tilde{x} + \epsilon)$ for all $\epsilon \in]0, \tilde{\epsilon}[$. It is not ruled out that $e' = e$ will hold again for larger values. For an illustration of the notation see Figure A-7.

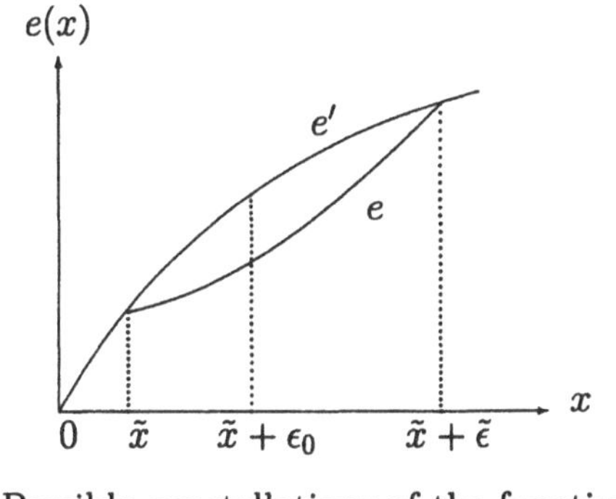

Figure A-7: Possible constellations of the functions $e(\cdot)$ and $e'(\cdot)$.

(i) First show that there exists an $\epsilon_0 \in]0, \tilde{\epsilon}[$ with:

$$e'(\tilde{x} + \epsilon_0) - e'(x) > e(\tilde{x} + \epsilon_0) - e(x) \quad \forall\, x < \tilde{x} + \epsilon_0. \tag{1}$$

This holds naturally for $x \leq \tilde{x}$, since for these values $e'(x) = e(x)$ and $e'(\tilde{x} + \epsilon_0) > e(\tilde{x} + \epsilon_0)$. It has to be shown, however, for $x \in]\tilde{x}, \tilde{x} + \epsilon_0[$. (1) is then equivalent to

$$e'(\tilde{x} + \epsilon_0) - e(\tilde{x} + \epsilon_0) > e'(x) - e(x) \quad \forall\, x \in]\tilde{x}, \tilde{x} + \epsilon_0[.$$

$$\Leftrightarrow e'(\tilde{x} + \epsilon_0) - e(\tilde{x} + \epsilon_0) > e'(\tilde{x} + \lambda\epsilon_0) - e(\tilde{x} + \lambda\epsilon_0) \; \forall \lambda \in]0, 1[. \tag{2}$$

By considering a new function that is given by the difference of the functions $e'(\cdot)$ and $e(\cdot)$, it is now shown that an ϵ–neighborhood of $\tilde{x}$ can be found in which (2) holds. Define

$$d(x) := e'(x) - e(x). \tag{3}$$

It then holds that

$$\begin{aligned} d(\tilde{x}) &= e'(\tilde{x}) - e(\tilde{x}) = 0, \\ d(\tilde{x}+\epsilon) &= e'(\tilde{x}+\epsilon) - e(\tilde{x}+\epsilon) > 0 \quad \forall\, \epsilon \in]0, \tilde{\epsilon}[. \end{aligned}$$

In order to show (2), i.e. that

$$\exists\, \epsilon_0 \in]0, \tilde{\epsilon}[: \quad d(\tilde{x}+\epsilon_0) > d(\tilde{x}+\lambda\epsilon_0) \quad \forall\, \lambda \in]0,1[,$$

I assume to the contrary that

$$\forall\, \epsilon \in]0, \tilde{\epsilon}[\quad \exists\, \lambda_\epsilon \in]0,1[: \quad d(\tilde{x}+\epsilon) \leq d(\tilde{x}+\lambda_\epsilon\epsilon). \tag{4}$$

Define $\epsilon^{(1)} := \lambda_\epsilon \epsilon$. Obviously $\epsilon^{(1)} \in]0, \tilde{\epsilon}[$ and therefore by assumption $\exists\, \lambda_{\epsilon^{(1)}} \in]0,1[$ with

$$\begin{aligned} d(\tilde{x}+\epsilon) &\leq d(\tilde{x}+\lambda_\epsilon\epsilon) = d(\tilde{x}+\epsilon^{(1)}) \\ &\leq d(\tilde{x}+\lambda_{\epsilon^{(1)}}\epsilon^{(1)}) = d(\tilde{x}+\lambda_{\epsilon^{(1)}}\lambda_\epsilon\epsilon). \end{aligned}$$

The same argument can be applied again and again, which leads to

$$\begin{aligned} d(\tilde{x}+\epsilon) &\leq d(\tilde{x}+\lambda_\epsilon\epsilon) \\ &\leq \ldots \leq \underbrace{d(\tilde{x}+\underbrace{\lambda_{\epsilon^{(n)}}\cdots\lambda_\epsilon}_{\to 0}\epsilon)}_{\to d(\tilde{x})=0}. \end{aligned}$$

As the number (n) of iterations approaches infinity, the product of the $\lambda_{\epsilon^{(i)}}$ converges to zero since they are element $]0,1[$. The two functions $e'(\cdot)$ and $e(\cdot)$ were assumed to be continuous, therefore the function $d(\cdot)$, being the difference of two continuous functions, is continuous as well. From the continuity of $d(\cdot)$ it then follows immediately that $\lim_{(\lambda_{\epsilon^{(n)}}\cdots\lambda_\epsilon\epsilon)\to 0} d(\tilde{x}+\lambda_{\epsilon^{(n)}}\cdots\lambda_\epsilon\epsilon) = d(\tilde{x}) = 0.$

However, $d(\tilde{x}+\epsilon) \leq 0$ contradicts $d(\tilde{x}+\epsilon) > 0$, which holds by the definition of $d(\cdot)$. Therefore there exists an $\epsilon_0 \in]0, \tilde{\epsilon}[$ with

$$\begin{aligned} d(\tilde{x}+\epsilon_0) &> d(\tilde{x}+\lambda\epsilon_0) \quad \forall\, \lambda \in]0,1[\\ \Leftrightarrow e'(\tilde{x}+\epsilon_0) - e(\tilde{x}+\epsilon_0) &> e'(\tilde{x}+\lambda\epsilon_0) - e(\tilde{x}+\lambda\epsilon_0) \\ \Leftrightarrow e'(\tilde{x}+\epsilon_0) - e'(\tilde{x}+\lambda\epsilon_0) &> e(\tilde{x}+\epsilon_0) - e(\tilde{x}+\lambda\epsilon_0) \\ \Leftrightarrow e'(\tilde{x}+\epsilon_0) - e'(x) &> e(\tilde{x}+\epsilon_0) - e(x) \end{aligned}$$

for all $x \in]\tilde{x}, \tilde{x}+\epsilon_0[$ and from above for all $x \in [0, \tilde{x}+\epsilon_0[$.

(ii) This result is now used to finalize the proof. Let $\tilde{\theta} := \tilde{x}+\epsilon_0$, i.e.

$$e'(\tilde{\theta}) - e'(\theta) > e(\tilde{\theta}) - e(\theta) \geq 0 \quad \forall\, \theta < \tilde{\theta}.$$

Since $f' \equiv f$, it follows

$$\frac{e'(\tilde{\theta}) - e'(\theta)}{f'(\tilde{\theta},\theta)} > \frac{e(\tilde{\theta}) - e(\theta)}{f(\tilde{\theta},\theta)} \quad \forall\, \theta < \tilde{\theta}.$$

Since this holds for every $\theta < \tilde{\theta}$, it holds in particular for the supremum of each of the expressions if $\arg\sup_{\theta<\tilde{\theta}} \frac{e(\tilde{\theta})-e(\theta)}{f(\tilde{\theta},\theta)} < \tilde{\theta}$. From $(e', a', f') \in F_I$ and the assumption that $a(\hat{\theta}) = \sup_{\theta<\hat{\theta}} \frac{e(\hat{\theta})-e(\theta)}{f(\hat{\theta},\theta)}$ for all $\hat{\theta}$ it then follows that $a'(\tilde{\theta}) > a(\tilde{\theta})$:

$$a'(\tilde{\theta}) \geq \sup_{\theta<\tilde{\theta}} \frac{e'(\tilde{\theta}) - e'(\theta)}{f'(\tilde{\theta},\theta)} > \sup_{\theta<\tilde{\theta}} \frac{e(\tilde{\theta}) - e(\theta)}{f(\tilde{\theta},\theta)} = a(\tilde{\theta}). \tag{5}$$

So, whenever $e'(\tilde{\theta}) > e(\tilde{\theta})$ for some $\tilde{\theta} \in [0, \theta_u]$ and $\arg\sup_{\theta<\tilde{\theta}} \frac{e(\tilde{\theta})-e(\theta)}{f(\tilde{\theta},\theta)} < \tilde{\theta}$, then $a'(\tilde{\theta}) > a(\tilde{\theta})$ for this $\tilde{\theta}$. Therefore, for a mechanism $(e, a, f) \in F_I$ with $a(\hat{\theta}) = \sup_{\theta<\hat{\theta}} \frac{e(\hat{\theta})-e(\theta)}{f(\hat{\theta},\theta)}$ for all $\hat{\theta}$, there exists no mechanism $(e', a', f') \in F_I$ with $e'(\cdot) \geq e(\cdot)$, $e'(\cdot) \neq e(\cdot)$ and $a'(\cdot) \leq a(\cdot)$. It follows that a mechanism $(e, a, f) \in F_I$ with $a(\hat{\theta}) = \sup_{\theta<\hat{\theta}} \frac{e(\hat{\theta})-e(\theta)}{f(\hat{\theta},\theta)}$ for all $\hat{\theta}$ is efficient in F_I.

3. In order to show that $a(\hat{\theta}) = \sup_{\theta<\hat{\theta}} \frac{e(\hat{\theta})-e(\theta)}{f(\hat{\theta},\theta)}$ for all $\hat{\theta}$ is sufficient for efficiency even if $\arg\sup_{\theta<\hat{\theta}} = \hat{\theta}$, it is now sufficient to show that whenever $e'(\hat{\theta}) > e(\hat{\theta})$ for some $\hat{\theta} \in [0, \theta_u]$, then (5) holds for at least one

$\tilde{\theta} \in [0, \theta_u]$. Reformulating (5) and using $\arg\sup_{\theta<\tilde{\theta}} = \tilde{\theta}$, $f' \equiv f$ and $f_\theta(\hat{\theta}, \theta) < 0$ yields

$$\lim_{\theta \to \tilde{\theta}} \left(\frac{e'(\tilde{\theta}) - e'(\theta)}{f(\tilde{\theta}, \theta)} - \frac{e(\tilde{\theta}) - e(\theta)}{f(\tilde{\theta}, \theta)} \right)$$

$$= \lim_{\theta \to \tilde{\theta}} \frac{(e'(\tilde{\theta}) - e(\tilde{\theta})) - (e'(\theta) - e(\theta))}{f(\tilde{\theta}, \theta)}$$

$$= \lim_{\theta \to \tilde{\theta}} \frac{-\frac{d}{d\theta} e'(\theta) + \frac{d}{d\theta} e(\theta)}{f_\theta(\tilde{\theta}, \theta)} \tag{6}$$

To finalize the proof it is now shown that $\frac{d}{d\theta} e'(\theta) > \frac{d}{d\theta} e(\theta)$ holds for some $\tilde{\theta} \in [0, \theta_u]$. Let, just as in the proof of part 2, $\tilde{x}$ be a value with $e'(\tilde{x}) = e(\tilde{x})$ (e.g. e'(0)=e(0)=0) and $e'(\tilde{x} + \epsilon) > e(\tilde{x} + \epsilon)$ for $\epsilon > 0$. If $e(\cdot)$ is constant in a neighborhood of $\tilde{x}$ and $e'(\cdot)$ is not, $\frac{d}{d\theta} e'(\theta) > \frac{d}{d\theta} e(\theta)$ holds trivially in this neighborhood. Let therefore $\frac{d}{d\theta} e(\theta) \neq 0$ (at least) in a neighborhood of $\tilde{x}$.

Application of the mean value theorem is then possible since $e'(\cdot)$ and $e(\cdot)$ are by assumption continuous in $\tilde{x} \leq \theta \leq \tilde{x} + \epsilon$ and differentiable in $\tilde{x} < \theta < \tilde{x} + \epsilon$. It follows that there exists at least one $\tilde{\theta}$ with

$$\frac{e'(\tilde{x} + \epsilon) - e'(\tilde{x})}{e(\tilde{x} + \epsilon) - e(\tilde{x})} = \frac{\frac{d}{d\theta} e'(\tilde{\theta})}{\frac{d}{d\theta} e(\tilde{\theta})}, \tag{7}$$

with $\tilde{x} < \tilde{\theta} < \tilde{x} + \epsilon$. The left hand side is larger than one and thus also the right hand side, $\frac{d}{d\theta} e'(\theta) > \frac{d}{d\theta} e(\theta)$ holds for some $\tilde{\theta} \in [0, \theta_u]$ and therefore, for this $\tilde{\theta}$,

$$\lim_{\theta \to \tilde{\theta}} \frac{-\frac{d}{d\theta} e'(\theta) + \frac{d}{d\theta} e(\theta)}{f_\theta(\tilde{\theta}, \theta)} > 0. \tag{8}$$

Plugging into (5) yields $a'(\tilde{\theta}) > a(\tilde{\theta})$. It follows that under the stated conditions a mechanism $(e, a, f) \in F_I$ with $a(\hat{\theta}) = \sup_{\theta<\hat{\theta}} \frac{e(\hat{\theta}) - e(\theta)}{f(\hat{\theta}, \theta)}$ for all $\hat{\theta}$ is efficient in F_I.

■

Appendix C

Proof of Corollary 1:
From (2.21) it holds that the audit probability in an efficient mechanism in F_I with proportional penalties is given by $a(\hat{\theta}) = \sup_{\theta<\hat{\theta}} \frac{e(\hat{\theta})-e(\theta)}{\gamma(\hat{\theta}-\theta)}$. The Corollary holds, if the supremum is attained for $\theta = 0$. This is the case, however, since regressivity implies that 1.) $\lim_{\theta\to 0} \frac{e(\hat{\theta})-e(\theta)}{\hat{\theta}-\theta} = \frac{e(\hat{\theta})}{\hat{\theta}} > \frac{d}{d\hat{\theta}}e(\hat{\theta}) = \lim_{\theta\to\hat{\theta}} \frac{e(\hat{\theta})-e(\theta)}{\hat{\theta}-\theta}$ and 2.) the supremum is not attained for any $0 < \tilde{\theta} < \hat{\theta}$. To see this assume the opposite. If $\arg\sup_{\theta<\hat{\theta}} \frac{e(\hat{\theta})-e(\theta)}{\hat{\theta}-\theta} = \tilde{\theta}$ for some $0 < \tilde{\theta} < \hat{\theta}$ then

$$\frac{e(\hat{\theta}) - e(\tilde{\theta})}{\hat{\theta} - \tilde{\theta}} > \frac{e(\hat{\theta})}{\hat{\theta}} \tag{9}$$

$$\Leftrightarrow \hat{\theta}[e(\hat{\theta}) - e(\tilde{\theta})] > (\hat{\theta} - \tilde{\theta})e(\hat{\theta}) \tag{10}$$

$$\Leftrightarrow \hat{\theta}e(\tilde{\theta}) < \tilde{\theta}e(\hat{\theta}) \tag{11}$$

$$\Leftrightarrow \frac{e(\tilde{\theta})}{\tilde{\theta}} < \frac{e(\hat{\theta})}{\hat{\theta}}, \tag{12}$$

which contradicts regressivity of $e(\cdot)$. ■

Proof of Proposition 3:
Part 1: To show that all mechanisms $(e, a, f_{prop}) \in F_I$ with $a(\hat{\theta}) = \frac{e(\hat{\theta})}{\gamma\hat{\theta}}$ have a regressive allocation function $e(\cdot)$, insert $a(\hat{\theta}) = \frac{e(\hat{\theta})}{\gamma\hat{\theta}}$ in the incentive compatibility condition (2.16). This yields

$$e(\theta) \geq e(\hat{\theta}) - \frac{e(\hat{\theta})}{\gamma\hat{\theta}} f(\hat{\theta}, \theta) \quad \forall\, \hat{\theta}, \theta. \tag{13}$$

Incentive compatibility holds for all $\hat{\theta} \leq \theta$, since $e(\cdot)$ is nondecreasing. Using

$f(\hat{\theta},\theta) = f_{prop}(\hat{\theta},\theta) = \max\{0, \gamma(\hat{\theta}-\theta)\}$ leads to

$$e(\theta) \geq e(\hat{\theta}) - \frac{e(\hat{\theta})}{\gamma\hat{\theta}}\gamma(\hat{\theta}-\theta) \qquad \forall\, \hat{\theta} > \theta \tag{14}$$

$$\Leftrightarrow \; e(\theta) \geq e(\hat{\theta}) - \frac{e(\hat{\theta})}{\hat{\theta}}\hat{\theta} + \frac{e(\hat{\theta})}{\hat{\theta}}\theta \quad \forall\, \hat{\theta} > \theta \tag{15}$$

$$\Leftrightarrow \; \frac{e(\theta)}{\theta} \geq \frac{e(\hat{\theta})}{\hat{\theta}} \qquad \forall\, \hat{\theta} > \theta, \tag{16}$$

which is a definition of regressivity.

Part 2: Inserting $a(\hat{\theta}) = \frac{1}{\gamma}D^{-}e(\hat{\theta})$ in the incentive compatibility condition as above yields for the proportional penalty

$$e(\theta) \geq e(\hat{\theta}) - \frac{1}{\gamma}D^{-}e(\hat{\theta})\gamma(\hat{\theta}-\theta) \qquad \forall\, \hat{\theta} > \theta, \tag{17}$$

which is a definition of convexity. To see this, remember that a convex function lies above the tangent in any point on the graph: $e(x+h) \geq e(x) + h\frac{d}{dx}e(x)$ with $h \neq 0$. Here $h < 0$ and we only consider the left-hand derivative in cases where the function is not differentiable. Substitute $\hat{\theta}$ for x and θ for $x+h$ (it follows $h = -(\hat{\theta}-\theta)$) to obtain (17). ∎

Proof of Proposition 5:
The proof is similar to the proof of Proposition 1, which is given in Appendix B.

"⇒" Suppose $(e, a, f_{prop}) \in F$ is efficient in F and $a(\hat{\theta}) = \frac{e(\hat{\theta})}{\gamma\hat{\theta}}$ does not hold. Then $a(\hat{\theta}) > \frac{e(\hat{\theta})}{\gamma\hat{\theta}}$ from (3.10) and it will be possible to decrease the audit probability without changing the allocations, which contradicts efficiency.

"⇐" Let $a(\hat{\theta}) = \frac{e(\hat{\theta})}{\gamma\hat{\theta}}$ for all $\hat{\theta} \in [0, \theta_u]$. Show that $(e, a, f_{prop}) \in F$ is efficient in F by contradiction: Let $(e', a', f_{prop}) \in F$ with $e'(\cdot) \geq e(\cdot)$, $e'(\cdot) \neq e(\cdot)$, i.e. $e'(\cdot) > e(\cdot)$ for some $\hat{\theta} \in [0, \theta_u]$. However, for this $\hat{\theta}$ it holds (from (3.10)):

$$a'(\hat{\theta}) \geq \frac{e'(\hat{\theta})}{\gamma\hat{\theta}} > \frac{e(\hat{\theta})}{\gamma\hat{\theta}} = a(\hat{\theta}).$$

Similar arguing holds if $a'(\cdot) \leq a(\cdot),\ a'(\cdot) \neq a(\cdot)$: For some $\hat{\theta} \in [0, \theta_u]$ it holds

$$e'(\hat{\theta}) \leq a'(\hat{\theta})\gamma\hat{\theta} < a(\hat{\theta})\gamma\hat{\theta} = e(\hat{\theta}).$$

Therefore there is no $(e', a', f_{prop}) \in F$ with (a) $e'(\cdot) \geq e(\cdot),\ e'(\cdot) \neq e(\cdot),\ a'(\cdot) \leq a(\cdot)$ or (b) $e'(\cdot) \geq e(\cdot),\ a'(\cdot) \leq a(\cdot),\ a'(\cdot) \neq a(\cdot)$ and $(e, a, f_{prop}) \in F$ with $a(\hat{\theta}) = \frac{e(\hat{\theta})}{\gamma\hat{\theta}}$ is efficient in F. ■

Proof of Proposition 9: 1. From (3.2) and $a(\hat{\theta}) = 1$ it follows that

$$1 = \sup_{\theta < \hat{\theta}} \frac{e(\hat{\theta}) - e(\theta)}{\gamma(\hat{\theta} - \theta)} \quad \forall\, \hat{\theta}. \tag{18}$$

If $e(\cdot)$ is linear, this only holds for the function $e(\hat{\theta}) = \gamma\hat{\theta}$. For $e(\cdot)$ concave, this implies $1 = \frac{e(\hat{\theta})}{\gamma\hat{\theta}} \Leftrightarrow e(\hat{\theta}) = \gamma\hat{\theta}$ and for $e(\cdot)$ convex $1 = \frac{1}{\gamma}D^{-}e(\hat{\theta}) \Leftrightarrow e(\hat{\theta}) = \gamma\hat{\theta}$ (no constant is added since $e(0) = 0$), cf. Proposition 2. If the allocation function is not purely concave or convex, these considerations can be applied to the single sections of the function, since (18) has to hold for each report $\hat{\theta}$.
2. From (3.19) it can be seen that $1 = \frac{e(\hat{\theta})}{C} \Leftrightarrow e(\hat{\theta}) = C$. ■

Appendix D

Proof of Proposition 12:

"$\Rightarrow$" is given in the text.

"$\Leftarrow$" Note that $f(\cdot)$ and l, and therefore $a(\cdot)$, are fixed. Suppose $(e, a_l, f) \in F$ is not efficient in F. Then exists a mechanism $(e', a_l, f) \in F$ with $e'(\cdot) \geq e(\cdot)$, $e'(\cdot) \neq e(\cdot)$. However, this contradicts feasibility (cf. (4.3)) and therefore (e, a_l, f) is efficient in F.

■

Proof of Corollary 10:
Let $F(l, \gamma, V, n) := \frac{V}{n\gamma} - \int_l^{\theta_u} (\theta - l) g(\theta) d\theta = 0$.

$$\begin{aligned}
\frac{\partial F}{\partial l} &= (l - l)g(l) + \int_l^{\theta_u} g(\theta) d\theta = G(\theta_u) - G(l) \\
&= 1 - G(l) > 0 \\
\frac{\partial F}{\partial \gamma} &= -\frac{V}{n\gamma^2} < 0 \\
\frac{\partial F}{\partial n} &= -\frac{V}{n^2\gamma} < 0 \\
\frac{\partial F}{\partial V} &= \frac{1}{n\gamma} > 0
\end{aligned}$$

Applying the implicit function theorem yields

$$\frac{\partial l}{\partial \gamma} > 0, \qquad \frac{\partial l}{\partial n} > 0, \qquad \frac{\partial l}{\partial v} < 0$$

■

Proof of Proposition 15:
With $TC_d = (1 - G(l)) \cdot n$ from (4.47) it follows

$$\frac{\partial TC_d}{\partial l^v} = -n \underbrace{G'(l^v)}_{>0} < 0,$$

$$\frac{\partial TC_d}{\partial \gamma} = \underbrace{\frac{\partial TC_d}{\partial l^v}}_{<0} \underbrace{\frac{\partial l^v}{\partial \gamma}}_{>0} < 0,$$

$$\frac{\partial TC_d}{\partial V} = \underbrace{\frac{\partial TC_d}{\partial l^v}}_{<0} \underbrace{\frac{\partial l^v}{\partial V}}_{<0} > 0.$$

■

Changes in Δ:

$$\frac{\Delta}{\gamma} = \underbrace{\frac{1}{\gamma^2}(\frac{\partial \lambda^v}{\partial \gamma}\gamma - \lambda^v)}_{<0} + \underbrace{\frac{dG}{dl^v}\frac{\partial l^v}{\partial \gamma}}_{>0}$$

$$\frac{\Delta}{V} = \underbrace{\frac{1}{\gamma}\frac{\partial \lambda^v}{\partial V}}_{>0} + \underbrace{\frac{dG}{dl^v}\frac{\partial l^v}{\partial V}}_{<0}$$

$$\frac{\Delta}{n} = \underbrace{\frac{1}{\gamma}\frac{\partial \lambda^v}{\partial n}}_{<0} + \underbrace{\frac{dG}{dl^v}\frac{\partial l^v}{\partial n}}_{>0}$$

■

Proof of Remark 12:

$$
\begin{aligned}
G(l) &= 1 - \frac{B}{n} \qquad \theta_l < l < \theta_u \\
&\Leftrightarrow \; \frac{B}{n} = \frac{(\frac{\theta_l}{l})^\kappa - (\frac{\theta_l}{\theta_u})^\kappa}{1 - (\frac{\theta_l}{\theta_u})^\kappa} \\
&\Leftrightarrow \; (\frac{\theta_l}{l})^\kappa = \frac{B}{n}[1 - (\frac{\theta_l}{\theta_u})^\kappa] + (\frac{\theta_l}{\theta_u})^\kappa \\
&\Leftrightarrow \; \frac{\theta_l}{l} = \left(\frac{B}{n}[1 - (\frac{\theta_l}{\theta_u})^\kappa] + (\frac{\theta_l}{\theta_u})^\kappa\right)^{\frac{1}{\kappa}} \\
\Leftrightarrow \quad l^* &= \theta_l \cdot \left(\frac{B}{n}[1 - (\frac{\theta_l}{\theta_u})^\kappa] + (\frac{\theta_l}{\theta_u})^\kappa\right)^{-\frac{1}{\kappa}} \\
&= \theta_l \cdot \left((\frac{\theta_l}{\theta_u})^\kappa[1 - \frac{B}{n}] + \frac{B}{n}\right)^{-\frac{1}{\kappa}}
\end{aligned}
$$

■

Calculations for Equation (4.70):

$$
\begin{aligned}
\frac{\partial l^*}{\partial B} &= \theta_l(-\frac{1}{\kappa})(\frac{B}{n}[1 - (\frac{\theta_l}{\theta_u})^\kappa] + (\frac{\theta_l}{\theta_u})^\kappa)^{-\frac{1+\kappa}{\kappa}} \frac{1}{n}[1 - (\frac{\theta_l}{\theta_u})^\kappa] < 0 \\
\frac{\partial l^*}{\partial n} &= \theta_l(-\frac{1}{\kappa})(\frac{B}{n}[1 - (\frac{\theta_l}{\theta_u})^\kappa] + (\frac{\theta_l}{\theta_u})^\kappa)^{-\frac{1+\kappa}{\kappa}} (-\frac{B}{n^2}[1 - (\frac{\theta_l}{\theta_u})^\kappa] > 0
\end{aligned}
$$

■

Proof of Remark 13:

$$
\begin{aligned}
\frac{\partial l^*}{\partial \theta_l} &= \left((\frac{\theta_l}{\theta_u})^\kappa[1-\frac{B}{n}]+\frac{B}{n}\right)^{-\frac{1}{\kappa}} \\
&\quad +\theta_l(-\frac{1}{\kappa})\left((\frac{\theta_l}{\theta_u})^\kappa[1-\frac{B}{n}]+\frac{B}{n}\right)^{-\frac{1+\kappa}{\kappa}}[1-\frac{B}{n}]\kappa(\frac{\theta_l}{\theta_u})^{\kappa-1}\frac{1}{\theta_u} \\
&= \left((\frac{\theta_l}{\theta_u})^\kappa[1-\frac{B}{n}]+\frac{B}{n}\right)^{-\frac{1}{\kappa}}\cdot\left[1-\underbrace{\frac{(\frac{\theta_l}{\theta_u})^\kappa[1-\frac{B}{n}]}{(\frac{\theta_l}{\theta_u})^\kappa[1-\frac{B}{n}]+\frac{B}{n}}}_{<1}\right] > 0
\end{aligned}
$$

$$
\frac{\partial l^*}{\partial \theta_u} = \left((\frac{\theta_l}{\theta_u})^\kappa[1-\frac{B}{n}]+\frac{B}{n}\right)^{-\frac{1+\kappa}{\kappa}}[1-\frac{B}{n}](\frac{\theta_l}{\theta_u})^{\kappa+1} > 0
$$

$$
\begin{aligned}
\frac{\partial l^*}{\partial \kappa} &= \theta_l\left((\frac{\theta_l}{\theta_u})^\kappa[1-\frac{B}{n}]+\frac{B}{n}\right)^{-\frac{1}{\kappa}}\cdot\ln\left(\underbrace{(\frac{\theta_l}{\theta_u})^\kappa[1-\frac{B}{n}]}_{<1-\frac{B}{n}}+\frac{B}{n}\right) \\
&\quad \cdot\frac{1}{\kappa^2}[1-\frac{B}{n}]\left((\frac{\theta_l}{\theta_u})^\kappa\underbrace{\ln(\frac{\theta_l}{\theta_u})}_{<0}\right) > 0
\end{aligned}
$$

■

Bibliography

Aban, I. B., M. M. Meerschaert and A. K. Panorska (2006): Parameter Estimation for the Truncated Pareto Distribution, *Journal of the American Statistical Association*, Vol. 101, No. 473, pp. 270–277.

Allingham, M. G. and A. Sandmo (1972): Income Tax Evasion: A Theoretical Analysis, *Journal of Public Economics*, Vol. 1, pp. 323–338.

Amacher, G. S., E. Koskela and M. Ollikainen (2005): Royalty Reform and Illegal Reporting of Harvest Volumes under Alternative Penalty Structures, paper presented at the 14th Annual Meeting of the European Association of Environmental and Resource Ecoonomists (EAERE), Bremen.

Arguedas, C. and H. Hamoudi (2004): Controlling Pollution with Relaxed Regulation, *Journal of Regulatory Economics*, Vol. 26, pp. 85–104.

Arrow, K. (1969): The Organization of Economic Activity: Issues Pertinent to the Choice of Market versus Nonmarket Allocation. The Analysis and Evaluation of Public Expenditures: The PPB System, Washington, DC: Joint Economic Committee, 91st Congress, 47-64.

Baron, D. and R. Myerson (1982): Regulating a monopolist with unknown costs, *Econometrica*, Vol. 50, pp. 911–930.

Baumol, W.J. and W.E. Oates (1971): The Use of Standards and Prices for Protection of the Environment, *Swedish Journal of Economics*, Vol. 73, pp. 42–54.

Becker, G. S. (1968): Crime and Punishment: An Economic Approach, *Journal of Political Economy*, Vol. 76, pp. 169–217.

Boadway, R. and M. Sato (2000): The optimality of punishing only the innocent: The case of tax evasion, *International Tax and Public Finance*, Vol. 7, pp. 641–664.

Border, K. and J. Sobel (1987): Samurai Accountant: A theory of Auditing and Plunder, *Review of Economic Studies*, Vol. 54, pp. 525–540.

Burroughs, S. M. and S. F. Tebbens (2001): Upper-truncated power laws in natural systems, *Journal of Pure and Applied Geophysics*, Vol. 158, pp. 741–757.

Burroughs, S. M. and S. F. Tebbens (2002): The upper-truncated power law applied to earthquake cumulative frequency-magnitude distortions, *Bulletin of the Seismological Society of America*, Vol. 92, pp. 2983–2993.

Burtraw, D. and K. Palmer (2004): SO_2 Cap-and-Trade Program in the United States: A "Living Legend" of Market Effectiveness, in: W. Harrington, R.D. Morgenstern and T. Sterner (Eds.), *Choosing Environmental Policy: Comparing Instruments and Outcomes in the United States and Europe*, Washinton, DC: Resources for the Future, pp. 41–66.

Chander, P. and L. L. Wilde (1998): A general Characterization of Optimal Income Tax Enforcement, *Review of Economic Studies*, Vol. 65, pp. 165–183.

Coase, R. (1960): The Problem of Social Cost, *Journal of Law and Economics*, Vol. 3, pp. 1–44.

Cohen, M.A. (2000): Monitoring and enforcement of environmental policy, in: H. Folmer and T. Tietenberg (Eds.), *The International Yearbook of Environmental and Resource Economics 1999/2000*, Edward Elgar, pp. 44–106.

Cramton, P. and S. Kerr (2002): Tradeable Carbon Permit Auctions: How and why to auction not grandfather, *Energy Policy*, Vol. 30, pp. 333–345.

Cremer, H. and F. Gahvari (1996): Tax Evasion and Optimal General Income Tax, *Journal of Public Economics*, Vol. 60, pp. 235–249.

Cremer, H., M. Marchand and P. Pestiau (1990): Evading, Auditing and Taxing: The Equity-Compliance Tradeoff, *Journal of Public Economics*, Vol. 43, pp. 67–92.

Crocker, T.D. (1966): The Structuring of Atmospheric Pollution Control Systems, in: H. Wolozin (Ed.), *The Economics of Air Pollution*, W.W.Norton& Co.

Cumberland, J. (1990): Public choice and the improvement of policy instruments for environmental management, *Ecological Economics*, Vol. 2, No. 2, pp. 149–162.

Dales, J.H. (1968): *Pollution, property and prices*, Toronto: University of Toronto Press.

Dasgupta, P., P. Hammond and E. Maskin (1979): The Implementation of Social Choice Rules: Some General Results on Incentive Compatibility, *Review of Economic Studies*, Vol. 46, pp. 185–216.

DEHSt (2005): Implementing the EU Emissions Trading System in Germany, Deutsche Emissionshandelsstelle, Slides presented by E. Harders at the Carbon EXPO, Cologne.

Dye, R. A. (1986): Optimal monitoring policies in agencies, *RAND Journal of Economics*, Vol. 17, pp. 339–350.

EC (2000): Green Paper on greenhouse gas emissions trading within the European Union (presented by the Commission), Brussels, European Commission.

EC (2003): Directive 2003/87/EC of the European Parliament and of the Council of 13 October 2003 establishing a scheme for greenhouse gas emission allowance trading within the Community and amending Council Directive 96/61/EC, Official Journal L 275, 25/10/2003, P.0032 - 0046.

EC (2004): Commission Decision of 29/01/04 establishing guidelines for the monitoring and reporting of greenhouse gas emissions pursuant to Directive 2003/87/EC of the European Parliament and of the Council, European Commission, Brussels: EC, *http : //www.environment – agency.gov.uk/commondata/*105385*/mrdecision*0104_775902*.pdf*.

EC and Ecofys (2007): Small Installations within the EU Emissions Trading Scheme, Report under the Project "Review of the EU Emissions Trading Scheme", European Commission Directorate General for Environment and Ecofys, Brussels: EC.

Egenhofer, C. and N. Fujiwara (2005): Reviewing the EU Emissions Trading Scheme - Priorities for short-term implementation of the second round of allocation: Part 1, CEPS Task Force Report No. 56, Brussels, Centre for European Policy Studies.

Gangadharan, L. (2000): Transaction Costs in Pollution Markets: An Empirical Study, *Land Economics*, Vol. 76, No. 4, pp. 601–614.

Gibbard, A. (1973): Manipulation of Voting Schemes: A General Result, *Econometrica*, Vol. 41, pp. 587–601.

Graichen, P. and T. Requate (2003): Der steinige Weg von der Theorie in die Praxis des Emissionshandels: Die EU-Richtlinie zum CO_2-Emissionshandel und ihre nationale Umsetzung, Christian-Albrechts-Universität Kiel, Economics Working Paper No 2003-08.

Green, J. and J.-J. Laffont (1986): Partially Verifiable Information and Mechanism Design, *Review of Economic Studies*, Vol. 53, pp. 447–456.

Groisman, P. Y., R. W. Knight, T. R. Karl, D. R. Easterling, B. Sun and J. H. Lawrimore (2004): Contemporary Changes of the Hydrological Cycle over the Contiguous United States: Trends Derived from In Situ Observations, *Journal of Hydrometeorology*, Vol. 5, No. 1, pp. 64–85.

Hahn, R. (1984): Market Power and Transferable Property Rights, *Quarterly Journal of Economics*, Vol. 99, pp. 753–765.

Hahn, R. (1989): Economic prescriptions for environmental problems, *Journal of Economic Perspectives*, Vol. 3, pp. 95–115.

Hahn, R.W. and G.L. Hester (1989b): Where did all the markets go? An analysis of EPA's Emissions Trading Program, *Yale Journal on Regulation*, Vol. 6, No. 1, pp. 109–153.

Hanley, N., S. Hallett and I. Moffatt (1990): Why is more notice not taken of economists' prescriptions for the control of pollution?, *Environment and Planning*, Vol. 22, pp. 1421 – 1439.

Hanley, N., J.F. Shogren and B. White (2007): *Environmental Economics in Theory and Practice, 2nd Edition*, New York: Palgrave MacMillan.

Harford, J.D. (1987): Self-Reporting of Pollution and the Firms' Behavior under Imperfectly Enforceable Regulations, *Journal of Environmental Economics and Management*, Vol. 14, pp. 293–303.

Harris, M and R. Townsend (1981): Resource Allocation Under Asymmetric Information, *Econometrica*, Vol. 49, pp. 33–64.

Harrison, D. and D. B. Radov (2002): *Evaluation of Alternative Initial Allocation Mechanisms in a European Union Greenhouse Gas Emissions Allowance Trading Scheme*, prepared for DG Environment, European Commission, Brussels.
http://europa.eu.int/comm/environment/climat/allocation_xsum.pdf.

Heyes, A.G. (2000): Implementing Environmental Regulation: Enforcement and Compliance, *Journal of Regulatory Economics*, Vol. 17, pp. 107–129.

Hurwicz, L. (1973): The Design of Mechanisms for Resource Allocation, *American Economic Review*, Vol. 63, pp. 1–30.

IPCC (2007): *Climate Change 2007*, Cambridge: Cambridge University Press.

Jensen, J. and T.N. Rasmussen (2000): Allocation of CO_2 Emissions Permits: A General Equilibrium Analysis of Policy Instruments, *Journal of Environmental Economics and Management*, Vol. 40, pp. 111–136.

Joskow, P.L. and R. Schmalensee (1998): The Political Economy of Marked-Based Environmental Policy: The U.S. Acid Rain Program, *Journal of Law and Economics*, Vol. 41, No. 1, pp. 37–83.

Kaplow, L. and S. Shavell (1994): Optimal Law Enforcement with Self-Reporting of Behavior, *Journal of Political Economy*, Vol. 102, No. 3, pp. 583–606.

Kerr, S. and D. Maré (1998): Transaction Costs and Tradeable Permit Markets: The United States Lead Phasedown, unpublished manuscript.

Kreps, D. M. (1990): *A Course in Microeconomic Theory*, New York: Harvester Wheatsheaf.

Krueger, A. O. (1974): The Political Economy of the Rent-Seeking Society, *American Economic Review*, Vol. 64, No. 3, pp. 291–303.

Kruger, J. and C. Egenhofer (2006): Confidence through Compliance in Emissions Trading Markets, CEPS Policy Brief No. 99/April 2006, Brussels, Centre for European Policy Studies.

Kruger, J. and W. A. Pizer (2004): The EU Emissions Trading Directive, Resources for the Future Discussion Paper.

Laffont, J.-J. and D. Martimort (2002): *The Theory of Incentives - The Principal-Agent Model*, Princeton, N.J.: Princeton University Press.

Liski, M. and J.-P. Montero (2005): A Note on Market Power in an Emission Permits Market with Banking, *Environmental and Resource Economics*, Vol. 31, pp. 159–173.

Lockard, A. and G. Tullock (2001): *Efficient Rent-Seeking: Chronicle of an Intellectual Quagmire*, Kluwer Academic Publishers.

Malamud, B. D., G. Morein and D. L. Turcotte (1998): Forest fires: an example of self-organized critical behavior, *Science*, Vol. 281, pp. 1840–1842.

Malik, A.S. (1993): Self-Reporting and the Design of Policies for Regulating Stochastic Pollution, *Journal of Environmental Economics and Management*, Vol. 24, pp. 241–257.

Mandelbrot, B. (1960): The Pareto-Lévy law and the distribution of income, *International Economic Review*, Vol. 1, No. 2, pp. 79–106.

Marhuenda, F. and I. Ortuno-Ortín (1997): Tax Enforcement Problems, *Scandinavian Journal of Economics*, Vol. 99, No. 1, pp. 61–72.

Mas-Colell, A., M. D. Whinston and J. R. Green (1995): *Microeconomic Theory*, Oxford: Oxford University Press.

McKibbin, W.J. and P.J. Wilcoxen (2002): The Role of Economics in Climate Change Policy, *Journal of Economic Perspectives*, Vol. 16, No. 2, pp. 107–129.

McKinsey and Ecofys (2005): Review of the EU Emissions Trading Scheme: survey highlights, DG Environment, Brussels: European Commission.

Michaelowa, A. (2004): Clean Development Mechanism and Joint Implementation, in: M. Lucht and G. Spangardt (Eds.), *Emissionshandel*, Springer, Heidelberg, pp. 137–152.

Mirrlees, J.A. (1971): An Exploration in the Theory of Optimum Income Taxation, *Review of Economic Studies*, Vol. 38, No. 2, pp. 175–208.

Montero, J.-P. (1997): Marketable pollution permits with uncertainty and transaction costs, *Resource and Energy Economics*, Vol. 20, pp. 27–50.

Montero, J.-P. (2008): A Simple Auction Mechanism for the Optimal Allocation of the Commons, *American Economic Review*, Vol. 98, No. 1, pp. 496–518.

Montgomery, W.D. (1972): Markets in Licenses and Efficient Pollution Control Programs, *Journal of Economic Theory*, Vol. 5, No. 3, pp. 395–418.

Mookherjee, D. and I. Png (1989): Optimal Auditing, Insurance and Redistribution, *Quarterly Journal of Economics*, Vol. 104, pp. 399–415.

Myerson, R. B. (1979): Incentive Compatibility and the Bargaining Problem, *Econometrica*, Vol. 47, pp. 61–73.

Myerson, R. B. (1981): Optimal Auction Design, *Mathematics of Operations Research*, Vol. 6, pp. 58–73.

NCEE (2004): *International Experiences with Economic Incentives for Protecting the Environment*, Washington, DC: Environmemntal Protection Agency.

Nordhaus (1993): Reflections on the Economics of Climate Change, *Journal of Economic Perspectives*, Vol. 7, No. 4, pp. 11–25.

Oreskes, N. (2004): Beyond the ivory tower - The scientific consensus on climate change, *Science*, Vol. 306, p. 1686.

Pareto, V. (1897): *Cours d'Economie Politique*, Lausanne and Paris: Rouge and Cie.

Polinsky, A.M. and S. Shavell (2000): The Economic Theory of Public Enforcement of Law, *Journal of Economic Literature*, Vol. 38, pp. 45–76.

Reilly, J., M. Mayer and J. Harnisch (2000): Multiple Gas Control Under the Kyoto Agreement, MIT Joint Program on the Science and Policy of Global Change, Report No. 58.

Reinganum, J. F. and L. L. Wilde (1985): Income Tax Compliance in a Principal-Agent Framework, *Journal of Public Economics*, Vol. 26, pp. 1–18.

Sandmo, A. (2005): The Theory of Tax Evasion: A Retrospective View, *National Tax Journey*, Vol. 58, No. 4, pp. 643 – 663.

Schmalensee, R., P.L. Joskow, A.D. Ellerman, J.P. Montero and E.M. Bailey (1998): An Interim Evaluation of Sulfur Dioxide Emissions Trading, *Journal of Economic Perspectives*, Vol. 12, No. 3, pp. 53–68.

Sharkey, W. W. (1982): *The theory of natural monopoly*, Cambridge, U.K.: Cambridge University Press.

Stavins, R.N. (1995): Transaction Costs and Tradeable Permits, *Journal of Environmental Economics and Management*, Vol. 29, pp. 133–148.

Stavins, R.N. (1998): What can we learn from the grand policy experiment? Lessons from SO_2 allowances trading, *Journal of Economic Perspectives*, Vol. 12, No. 3, pp. 69–88.

Stavins, R.N. (2001): Lessons from the American Experiment with Market-Based Environmental Policies, KSG Faculty Research Working Paper Series, Harvard University.

Stern, N. (2006): *The Economics of Climate Change: The Stern Review*, Cambridge: Cambridge University Press.

Sterner, T. and U.M. Persson (2007): An even Sterner Review, resources for the Future Discussion Paper, Washington, RFF DP 07-37.

Tietenberg, T.H. (2006): *Emissions trading: principles and practice - 2nd ed.*, Washington, DC: Resources for the Future.

Tullock, G. (1967): The Welfare Costs of Tariffs, Monopolies, and Theft, *Western Economic Journal*, Vol. 5, pp. 224–232.

Tullock, G. (1980): *Efficient Rent-Seeking*, College Station: Texas A&M University Press., pp. 97–112.

Umweltbundesamt (2004): Prüfungsrichtlinie zur Verifizierung von Zuteilungsanträgen, Deutsche Emissionshandelsstelle im Umweltbundesamt.

Umweltbundesamt (2005): Implementing Monitoring, Reporting and Verification for the EU ETS, slides presented by F. Schafhausen at the Carbon EXPO, Cologne.

UN (2001): *Greenhouse Gas Market Perspectives: Trade and Investment Implications of the Climate Change Regime*, United Nations, New York and Geneva.

UNFCCC (1997): *Kyoto Protocol to the United Nations Framework Convention on Climate Change*, FCCC/CP/L.7/Add.1, Kyoto.: United Nations Framework Convention on Climate Change.

Vylder, F. de, M. Goovaerts and J. Haezendonck (1984): *Premium Calculation in Insurance*, Kluwer Academic Publishers.

Yitzhaki, S. (1974): A note on income tax evasion: A theoretical analysis, *Journal of Public Economics*, Vol. 3, pp. 201–202.

Kollektive Entscheidungen, Wirtschaftspolitik und öffentliche Finanzen

Herausgegeben von Cay Folkers

Band 1 Gerald Pech: Besteuerung und Staatsverschuldung in der Demokratie. Zur dynamischen Analyse staatlicher Budgetpolitiken. 1996.

Band 2 Wolfgang Rippin: Theorie spekulativer Blasen. 1997.

Band 3 Claus Schnabel: Tariflohnpolitik und Effektivlohnfindung. Eine empirische und wirtschaftspolitische Analyse für die alten Bundesländer. 1997.

Band 4 Michael Grüne: Subventionen in der Demokratie. Analytische Grundlagen einer Subventionsordnung. 1997.

Band 5 Frank Ziegele: Hochschule und Finanzautonomie. Grundlagen und Anwendung einer politisch-ökonomischen Theorie der Hochschule. 1997. 2., durchges. Aufl. 1998.

Band 6 Petra Mühl-Schimmele: Die Behandlung von Erbschaften und Schenkungen in einem konsumorientierten Einkommensteuersystem. 1999.

Band 7 Stefan Boeters: Arbeitsmarktwirkungen einer budgetneutralen Reform der Besteuerung von Produktionsfaktoren. 2000.

Band 8 Frank Brocks: Die staatliche Förderung alternativer Kraftstoffe: Das Beispiel Biodiesel. 2001.

Band 9 Roman Dawid: Demokratische Kontrolle staatlicher Budgetierungsprozesse – Das Haushaltsverfahren der Europäischen Union. 2002.

Band 10 Karl Justus Bernhard Neumärker: Die politische Ökonomie der privaten Bereitstellung öffentlicher Güter. Ein vernachlässigtes Grundproblem der Finanzwissenschaft. 2003.

Band 11 Gerhard Schick: Doppelter Föderalismus in Europa. Eine verfassungsökonomische Untersuchung. 2003.

Band 12 Luis Gerardo González Morales: The Economics of Corruption and Bureaucratic Inefficiency in Weak States. Theory and Evidence. 2003.

Band 13 Stefan Kotte: Politische Transaktionskosten in der Demokratie. 2004.

Band 14 Olaf Schulemann: Horizontale Gerechtigkeit und Einkommensbesteuerung. 2008.

Band 15 Frauke Eckermann: Efficient Enforcement of Truth-Telling in the Grandfathering Process of an Emissions Trading Scheme. 2008.

Weitere Veröffentlichungen zum Themenkreis dieser Reihe

Claus Schnabel: Zur ökonomischen Analyse der Gewerkschaften in der Bundesrepublik Deutschland. Theoretische und empirische Untersuchungen von Mitgliederentwicklung, Verhalten und Einfluß auf wirtschaftliche Größen, 1989.

Thomas Märtz: Interessengruppen und Gruppeninteressen in der Demokratie. Zur Theorie des Rent-Seeking, 1990.

Peter Mendler: Zur ökonomischen und politisch-institutionellen Analyse öffentlicher Kredithilfen, 1992.

Karin Beckmann: Probleme der Regionalpolitik im Zuge der Vollendung des Europäischen Binnenmarktes. Eine ökonomische Analyse, 1995.

Karl Justus Bernhard Neumärker: Finanzverfassung und Staatsgewalt in der Demokratie. Ein Beitrag zur konstitutionellen Finanztheorie, 1995.

Peter Lang · Internationaler Verlag der Wissenschaften

Ignacio Pérez Domínguez

Greenhouse Gases: Inventories, Abatement Costs and Markets for Emission Permits in European Agriculture

A Modelling Approach

Frankfurt am Main, Berlin, Bern, Bruxelles, New York, Oxford, Wien, 2006.
213 pp., num. fig. and tab.
European University Studies: Series 5, Economics and Management. Vol. 3184
ISBN 978-3-631-55082-3 · pb. € 41.10*

The international community is putting some pressure on scientific researchers to come up with reliable indicators to estimate emissions from greenhouse gases. The development towards integrative multi gas strategy approaches has allowed the inclusion of the agricultural sector in the political agenda. Modelling alternatives for the estimation of emission factors, definition of policy instruments for greenhouse gas emission abatement as well as measurement of their economic effects are at this stage quite important for the coming multilateral negotiations.

Contents: Climate change and agriculture · Modelling of greenhouse gas emission inventories for the European agricultural sector · The use of emission abatement instruments: tradable permits as a viable option in agriculture · Calculation of marginal abatement cost curves · Modelling a market of tradable emission permits · Comparison between emission abatement instruments including price effects

FSC
www.fsc.org
MIX
Papier | Fördert
gute Waldnutzung
FSC® C083411

Kolibri 360 GmbH
Ferdinand-Jühlke-Straße 7
99095 Erfurt, Deutschland
produktsicherheit@kolibri360.de

Druck:
CPI Druckdienstleistungen GmbH
im Auftrag der
Kolibri 360 GmbH
Ferdinand-Jühlke-Str. 7
99095 Erfurt